Hepatitis C in Developing Coun

Contents

3.5   Hepatitis C in Developing Countries in Southeast Asia
Thi Q. Doan

Section 4
Hepatitis C Coinfections and Comorbidities in Developing Countries

4.1 Hepatitis C and Schistosomiasis Coinfection

Sanaa M. Kamal

4.2 Hepatitis C and Helminthic Infections

Khalifa S. Khalifa and Othman Amin

4.3 Hepatitis C and HIV Coinfection in Developing Countries

Ozlem Tastan Bishop

4.4 Hepatitis B and C Coinfection

Georgios Zacharakis

4.5 Hepatitis C Infection in Patients With Hemoglobinopathies

Sanaa M. Kamal and Ahmed M. Fouad

4.6 Hepatitis C in Patients on Hemodialysis

Tamer A. Hafez

Section 5
Evolution of Hepatitis C Treatment

List of Contributors

Ahmed Abdel Aziz King Abdullah University for Science and Technology, Jeddah, Kingdom of Saudi Arabia

Sara A. Abdelhakam Ain Shams University, Cairo, Egypt

Othman Amin Al Khartoum University, Khartoum, Sudan

Ozlem Tastan Bishop Rhodes University, Grahamstown, South Africa

Mohamed A. Daw University of Tripoli, Tripoli, Libya

Thi Q. Doan Hanoi Medical University, Ho Chi Minh City, Vietnam

Ahmed M. Fouad American University of Lebanon, Beirut, Lebanon

Huda H. Gaafar Prince Sattam College of Medicine, Kharj, Kingdom of Saudi Arabia

Dahlia Ghoraba Ain Shams Faculty of Medicine, Cairo, Egypt

Tamer A. Hafez American University, Cairo, Egypt

Sanaa M. Kamal Ain Shams Faculty of Medicine, Cairo, Egypt

Khalifa S. Khalifa Ain Shams Faculty of Medicine, Cairo, Egypt

Osi Obadahn University of Lagos, Lagos, Nigeria

Mohamad A. Othman University of Gezira, Khartoum, Sudan

Shashi Shekhar Dar Al Uloom University College of Medicine, Riyadh, Kingdom of Saudi Arabia

Georgios Zacharakis University of Nicosia, Nicosia, Republic of Cyprus; Prince Sattam bin Abdulaziz University, Al Kharj, Saudi Arabia

Introduction

Hepatitis C infection is a major public health issue worldwide [1–4]. The early stages of the infection are mostly asymptomatic and progress to chronic hepatitis, cirrhosis, or hepatocellular carcinoma (HCC). About 180 million people are infected worldwide with hepatitis C and most infected cases are in resource-limited countries [2,3]. The prevalence of hepatitis C virus (HCV) is much higher among at-risk populations (e.g., drug users, HIV-positive individuals, and prisoners) [4]. The deficient resources in such underprivileged regions pose significant barriers to containing the HCV epidemic in these countries. HCV prevalence can reach alarming levels, such as 60%–80% in injection drug users and more than 80% in HIV-coinfected individuals, particularly in developing countries where the number of injecting drug users is especially high [3,4]. The poor medical infrastructures, shortage of well-trained health-care workers, insufficient screening, abundance of risk factors for HCV transmission, and poor access to health care and treatment contribute to the huge burden of HCV in resource-limited countries [5]. Chronic HCV accounts for a substantial percentage of cirrhosis and HCC particularly in the endemic regions such as Egypt, sub-Saharan Africa, and South-East Asia.

The persistence and increased morbidity from HCV infection result from several factors including poor health education, poverty, inadequate prevention and screening strategies, frequent iatrogenic transmission, absence of insufficient medical insurance, prevalence of coinfection, stigma associated with the infection, and insufficient or even inexistent access to treatment [6]. In resource-limited countries, the inadequate infrastructures hamper the conduction of proper surveillance for collection of accurate data on the actual prevalence of HCV among the general population and vulnerable individuals. Despite the recent considerable strides in treatment and eradication of HCV in wealthy countries, reduction in transmission and morbidity from chronic HCV in poor countries remains a distant objective. Poverty and illiteracy are major factors that ignite HCV transmission in resource-limited countries. According to the World Health Organization, less than 50% of the blood supply in sub-Saharan Africa is screened for HCV, and 2–3 million people are infected each year through unsafe injections. Access to treatment is also a major obstacle to containing the epidemic in resource-limited countries.

The high burden of HCV in poor countries is not restricted to such regions. Recently, HCV spreads beyond its strongholds in Africa and South East Asia and posed a real threat to the health systems in Europe and North America. The recent mass migration waves introduced increasing numbers of documented or undocumented migrants or trafficked persons who come from countries with high HCV prevalence. Many migrants are not properly screened upon arrival to the host countries, and infected individuals may not have access to treatment. Furthermore, migrants in many countries are kept in camps where transmission of infections such as HIV, HBV, and HCV may be increased because of traditional and social behaviors. Thus, the increasing influx of refugees from highly endemic counties may change the disease burden in Western refugee-hosting countries. To date, WHO does not recommend obligatory screening of refugee and migrant populations for communicable diseases including HCV because there is no clear evidence of benefits (or cost-effectiveness); furthermore, it can cause anxiety in individual refugees and the wider community. Obligatory screening may deter migrants from asking for a medical check-up, thus jeopardizing identification of high-risk patients [7].

Taken together, the burden status of hepatitis C in resource-limited countries has been ignored and neglected for long intervals. Accurate estimates of the incidence and prevalence of HCV in developing countries are critical for designing effective strategies to combat the epidemic. Absence of a vaccine for HCV stresses the importance of strengthening preventive efforts and early identification and treatment of infected individuals. The advent of highly potent, new interferon-free antiviral drugs mandates serious collaborative efforts of health policy-makers, medical doctors, scientists, and governments to improve access to screening, care, and HCV treatment for patients living in resource-limited settings. To achieve this, it is crucial to conduct large clinical trials and national programs to adapt screening and treatment strategies to the local conditions of resource-poor settings, and cost-effectiveness studies should be carried out urgently as these are needed to demonstrate the feasibility, safety, and benefits of such therapeutic regimen for patients living in resource-poor HCV endemic countries.

This book explores the different aspects of the long-forgotten problem of hepatitis C infection in developing countries. The book chapters provide invaluable insight and up-to-date information about the HCV epidemiology and modes of transmission in different developing countries. The HCV genotypes, natural history, and treatment are analyzed. The impact of HCV on public health, economy, and social structure of affected countries is discussed.

Dr. Sanaa M. Kamal

REFERENCES

[1] Lavanchy D. Evolving epidemiology of hepatitis C virus. Clin Microbiol Infect 2011;17:107–15.

[2] World Health Organization. Hepatitis C. WHO fact sheet 164. Geneva, Switzerland: World Health Organization; 2000. Available from: http://www.who.int/mediacentre/factsheets/fs164/en/.

[3] Lavanchy D. The global burden of hepatitis C. Liver Int 2009;29(Suppl. 1):74–81.

[4] Shepard CW, Finelli L, Alter MJ. Global epidemiology of hepatitis C virus infection. Lancet Infect Dis 2005;5:558–67.

[5] Lemoine M, Thursz M. Hepatitis C, a global issue: access to care and new therapeutic and preventive approaches in resource-constrained areas. Semin Liver Dis 2014;34:89–97.

[6] Lemoine M, Eholie S, Lacombe K. Reducing the neglected burden of viral hepatitis in Africa: strategies for a global approach. J Hepatol 2015;62:469–76.

[7] Assessing the burden of key infectious diseases affecting migrant populations in the EU/EEA, European Centre of Disease Prevention and Control. 2014. [Internet] at: http://www.europarl.europa.eu/RegData/etudes/BRIE/2016/573908/EPRS_BRI(2016)573908_EN.pdf.

Section 1

Hepatitis C Virus: Virology, Epidemiology, and Transmission

Chapter 1.1

Hepatitis C Virus: Virology and Genotypes

Ahmed Abdel Aziz

King Abdullah University for Science and Technology, Jeddah, Kingdom of Saudi Arabia

Chapter Outline

INTRODUCTION

Hepatitis C virus (HCV) is a positive-strand RNA virus distantly related to the Flaviviridae family. The HCV genome consists of about 9400 nucleotides, with one large open reading frame encoding for a polypeptide about 3000 amino acids long, consisting of structural and nonstructural domains. The sequence contains a 5′ untranslated region (5′ UTR) of 341 bases, a long open reading frame coding for a polyprotein of 3011 amino acids, and a 3′ untranslated region of about 27 bases. The three N-terminal HCV proteins (C, E1, and E2/NS2) are probably structural, and the four C-terminal proteins (NS2, NS3, NS4, and NS5) are believed to function in viral replication [1,2] (Fig. 1.1.1).

The HCV lifecycle begins with the attachment of a virion to its specific receptors on hepatocytes [2]. The high-density lipoprotein receptor scavenger receptor class B type I, tetraspanin CD81, tight-junction protein claudin-1, and occludin are the known cellular receptors initiating the attachment step of HCV infection. After binding with its receptor complex, the viron is internalized, and

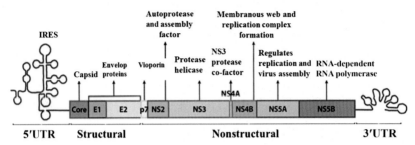

FIGURE 1.1.1   Structure of hepatitis C virus.

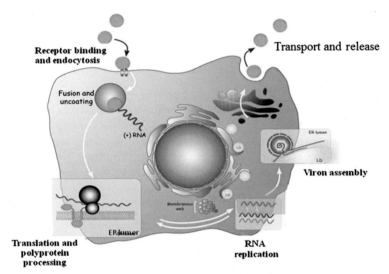

FIGURE 1.1.2   Life cycle of hepatitis C virus.

the nucleocapsid is released into the cytoplasm. Uncoating then occurs with exposure of the genomic RNA, which is used both for polyprotein translation and replication in the cytoplasm. HCV replication takes place within the "replication complex" containing the viral nonstructural proteins and cellular proteins [1] (Fig. 1.1.2).

HCV replication is catalyzed by the NS5B protein and other viral nonstructural proteins. The NTPase/helicase domain of the NS3 protein is related to viral replication, including RNA-stimulated NTPase activity, RNA binding, and unwinding of RNA regions of extensive secondary structure. NS4B initiates the formation of a replication complex that supports HCV replication. The NS5A protein also plays an important regulatory role in virus replication. A number of cellular factors are involved in HCV replication, such as cyclophilin A, required for HCV replication through its interaction with NS5A and the NS5B, and microRNA-122, which helps HCV replication through its binding with the 5′

UTR of the HCV genome. Therefore, host factors may also become the potential targets for anti-HCV therapies. At present, at least two host-targeted agents have reached clinical development, including specific inhibitors of cyclophilin A and antagonists of microRNA-122 [1–4].

The genomes of HCV display significant sequence heterogeneity and have been classified into genotypes and subtypes. To date, six genotypes (1–6) have been recognized [2]. Genotypes differ from each other by 30%–35% of nucleotide sequences. More variability is concentrated in regions such as the E1 and E2 glycoproteins, whereas sequences of the core gene and some of the nonstructural protein genes, such as NS3, are more conserved. The 5′ UTR contains a relatively well-conserved region that has a 92% homology among different HCV genotypes [3]. Some genotypes are further divided into subtypes (a, b, c, etc.), which typically differ from each other by 20%–25% in nucleotide sequences. Within a given individual, multiple quasi-species can exist [4,5].

EPIDEMIOLOGY OF HEPATITIS C VIRUS GENOTYPES AND GENOTYPIC VARIATIONS AROUND THE WORLD

HCV is divided into six distinct genotypes throughout the world with multiple subtypes in each genotype class. A genotype is a classification of HCV virus based on the genetic material in the RNA strands (Fig. 1.1.3).

Genotypes 1, 2, and 3 (HCV G1, G2, and G3) have a worldwide distribution [5,6]. Types HCV G1a and G1b predominate in Europe, North and South America, and Japan, accounting for about 60% of global infections. HCV G2 is common in Japan and China and is found in the United States and in northern, western, and southern Europe [6,7]. HCV G3 is prevalent in Southeast Asia,

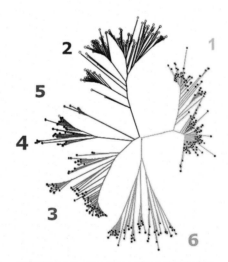

FIGURE 1.1.3   Genotypes of hepatitis C virus.

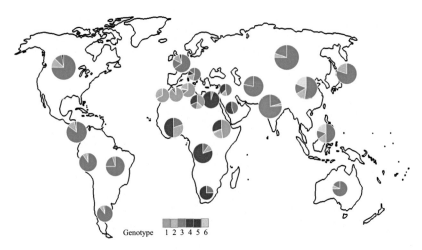

FIGURE 1.1.4 Geographical distribution of hepatitis C virus genotypes.

India, and Australia and is variably distributed in Europe and the United States [5,6]. HCV G4 is principally found in the Middle East, Egypt, and central Africa [8,9]. HCV G5 is almost restricted to South Africa [10], and HCV G6 is found in specific regions in Asia [11,12] (Fig. 1.1.4).

The origin and evolution of HCV genotypes in humans is still not clear. It is believed that the virus has evolved over several thousand years [1]. The huge genetic diversity of genotypes HCV G1, G2, and G4 in western and central Africa [12–17] and of HCV genotypes G3 and G6 in the Indian subcontinent and South East Asia [11,12] suggests that HCV has been endemic in these populations for a long time. The appearance of new risk groups and routes of transmission such as blood transfusion, the medical use of unsterilized needles, and (especially in industrialized countries) injection drug use and the sharing of injection equipment allowed the rapid spread and global distribution of genotypes HCV G1, G2, and G3 to Western and other nontropical countries over the past 30–70 years [15]. However, a close association between specific genotypes and specific geographic regions is observed, such as the prevalence of HCV G4a in Egypt [13–15] and HCV G5a in South Africa [10]. In Western countries, specific genotypes circulate in various risk groups, including HCV G1a and G3a in injecting drug users (IDUs) and HCV G1b, G2a, G2b, G2c, and G4a throughout the Mediterranean area [13,15].

Over the past decade, the epidemiology and distribution of HCV genotypes have shown marked variations in some geographic areas and have been relatively stable in others. Much of the HCV genotype variability detected between geographic regions can be explained by the different HCV epidemic histories in these regions, the frequency and extent to which different risk factors have contributed to the transmission of HCV, and population dynamics, which directly or indirectly introduced specific genotypes and their hosts into new environments and promoted the circulation of genotypes different from HCV G1, G2, and G3 [18].

HEPATITIS C VIRUS GENOTYPE PATTERNS IN NORTH AFRICA AND THE MIDDLE EAST

The genotype distribution in the Middle East is characterized by two main patterns: the prevalence of HCV G4 in Arab countries in the eastern regions of the Middle East, including Egypt [8,13–15], Saudi Arabia [9], Syria, and the United Arab Emirates [13,14], and the predominance of genotype HCV G1a or G1b in western areas of North Africa such as Tunisia and Morocco [19] and in non-Arab countries such as Turkey and Iran [20]. Egypt has the highest prevalence of HCV worldwide (13%) [18] and almost 90% of those infected with HCV in this country have HCV G4 [13,14]. The current sequence diversity and phylogenetic tree structure of HCV G4a in Egypt are compatible with the introduction of HCV into that population through parenteral treatment for schistosomiasis with indispensable and poorly sterilized needles in the 1950s and 1960s [13]. Although subtype HCV G4a is the dominant Egyptian HCV strain, recent studies revealed that other subtypes are also present, indicating that HCV G4 in Egypt is extremely variable not only in terms of sequence but also in terms of functional and immunologic determinants [22,23]. In Saudi Arabia, HCV G4 constitutes 62% of HCV infection, followed by HCV G1 (24%), HCV G3 (9%), and HCV G5 (0.3%). Unlike the predominance of subtype 4a in Egypt, however, subtypes HCV G4c and G4d are the most prevalent in Saudi Arabia, suggesting that the origin and transmission of HCV G4 in that country might be different from those in Egypt [9].

HEPATITIS C VIRUS GENOTYPE PATTERNS IN SUB-SAHARAN AFRICA

HCV is highly prevalent in Africa, with more than 30 million individuals infected [12,13,21]. The distribution of HCV genotypes in sub-Saharan countries is less well documented than in other geographic areas. Infections in West Africa are caused predominantly by HCV G2 [13], whereas those in central Africa, such as the Democratic Republic of Congo and Gabon, are caused by genotypes HCV G1 and G4 [17,21]. In both regions, genotypes HCV G1, G2, and G4 are highly heterogeneous, with 20–30 subtypes detected in some regions. This finding indicates a unique pattern of long-term HCV infection that has not been observed elsewhere. It has been thus hypothesized that these HCV genotypes originated and diversified for long periods in West or Central Africa before spreading to other regions [13,16,17,21]. The high population density in Africa, population shift, migration, and poor socioeconomic conditions—and importantly the high adaptation of HCV to its human host—ensured the long persistence of HCV infection and its successful ongoing transmission.

HCV G5 is the predominant genotype in South Africa (40%), followed by HCV G1 (33%), HCV G2 and G3 (21.5%), and HCV G4 (2.3%) [10,21]. A phase of exponential spread in South Africa was characterized by high growth rates indicative of rapid epidemic spread. A slowdown was observed around 1950, but this does not appear to be consistent with transmission through blood products, which continued until 1990. The South African dynamics might perhaps reflect an iatrogenic intervention in the early 20th century, such as the use of unsterilized medical equipment [10].

HEPATITIS C VIRUS GENOTYPE PATTERNS IN SOUTHEAST ASIA

Southeast Asia is characterized by a high prevalence of HCV, with more than 32 million individuals infected. Genotypes HCV G3, G1, G6, and G2 are prevalent in this area [1,2]. The high prevalence of HCV and the genotypic diversity in Southeast Asia have important epidemiologic implications given the increasing numbers of Asian immigrants in Europe and the United States and the dynamic population mobility in Asia.

Genotypes HCV G1, G2, G3, and G6 are found in China, with different geographic and demographic distributions. HCV G1b is the most widely distributed, and HCV G3 and G6 are more prevalent in southwest China [24–26]. In Taiwan, HCV subtypes G1b and G2a are the major subtypes. However, HCV G6 is found in south China, Hong Kong, Taiwan, Singapore, Malaysia, Vietnam, and Thailand. The main genotypes in Japan and Korea are HCV G1 and G2 [27,28]. In the Philippines, HCV G3 is predominant (35%), followed by HCV G1a (35%), HCV G1b (15%), and HCV G2 (7%) [28]. HCV G3 is the predominant genotype (55%) in this geographic region followed by HCV G1 (25%) and HCV G2 (20%) [29,30].

HEPATITIS C VIRUS GENOTYPE DISTRIBUTION IN EASTERN EUROPE

Across much of Europe, HCV G1 is the most commonly encountered variant, followed by genotypes HCV G2 and G3 [31]. Recently, genotypes HCV G3a and G4 have been frequently detected among IDUs in Europe [31]. However, viral genetic sequences have become more heterogeneous over time in several European countries. Recently, a shift was observed in the Netherlands and Belgium from a dominant prevalence of HCV G1 to an increasing prevalence of non-G1 genotypes, particularly among IDUs [32]. Immigration and IDU introduced HCV genotypes that have not been common in Europe. Recently, however, HCV G1b is progressively being replaced by HCV G3a, particularly among IDUs [33].

In Russia and most Eastern European countries, GT1 has been reported to account for the vast majority of HCV infections with prevalence at 75%. HCV G1 is followed by HCV G3a, G2a, and G1a. HCV genotypes and subtypes in

Europe have become more heterogeneous over time probably through two separate epidemics of HCV during the past 50 years. The first involved the spread of genotypes HCV G1, G2, and G3. The second was more recent and involved the introduction of specific genotypes such as HCV G1a, G3a, and G4 into new risk groups by the movement of IDUs across European borders [34].

HEPATITIS C VIRUS GENOTYPE PATTERNS IN THE SOUTH AMERICA

The HCV genotype distribution in South America is similar to that in North America, with prevalence of genotypes HCV G1, G3, and G2 in Brazil, Argentina, and Peru [35,36]. The genotype distribution in South America is relatively stable, with less genotype diversity than in other geographic areas, a finding that might be explained by relatively limited population dynamics in South America or less interaction through immigration or travel with populations from countries with diverse HCV genotypes.

CONCLUSIONS

The distribution of HCV genotypes is changing worldwide. Genotypes that have been restricted to certain geographic areas have been spreading elsewhere. Population dynamics, including population density, subdivision, and migration, play an important role in altering the HCV genotype distribution in various countries and risk groups. Knowledge of the genotype distribution in specific developing countries is critical to determine the cost-effective treatment strategies that provide the maximum sustained virologic response.

ABBREVIATIONS

HCV Hepatitis C virus
G1 Genotype 1
G2 Genotype 2
G3 Genotype 3
G4 Genotype 4
G5 Genotype 5
G6 Genotype 6

REFERENCES

[1] Simmonds P. The origin of hepatitis C virus. Curr Top Microbiol Immunol 2013;369:1–15.
[2] Choo QL, Kuo G, Weiner AG, Overby LR, Bradley DW, Houghton M. Isolation of a cDNA clone derived from a blood-borne non-A, non-B viral hepatitis genome. Science 1989;244:359–62.
[3] Santolini E, Migliaccio G, La Monica N. Biosynthesis and biochemical properties of the hepatitis C virus core protein. J Virol 1994;68:3631–41.

[4] Goffard A, Callens N, Bartosch B, Wychowski C, Cosset FL, Montpellier C, Dubuisson J. Role of N-linked glycans in the functions of hepatitis C virus envelope glycoproteins. J Virol 2005;79:8400–9.

[5] Simmonds P. Viral heterogeneity of the hepatitis C virus. J Hepatol 1999;31:54–60.

[6] Simmonds P. Genetic diversity and evolution of hepatitis C virus—15 years on [review]. J Gen Virol 2004;85:3173–88.

[7] Simmonds P, Bukh J, Combet C, et al. Consensus proposals for a unified system of nomenclature of hepatitis C virus genotypes. Hepatology 2005;42:962–73.

[8] Abdel-Aziz F, Habib M, Mohamed MK, et al. Hepatitis C virus (HCV) infection in a community in the Nile Delta: population description and HCV prevalence. Hepatology 2000;32:111–5.

[9] Shobokshi OA, Serebour FE, Skakni L, et al. Hepatitis C genotypes and subtypes in Saudi Arabia. J Med Virol 1999;58:44–8.

[10] Smuts HE, Kannemeyer J. Genotyping of hepatitis C virus in South Africa. J Clin Microbiol 1995;33:1679–81.

[11] Wasitthankasem R, Vongpunsawad S, Siripon N, Suya C, Chulothok P, Chaiear K, Rujirojindakul P, Kanjana S, Theamboonlers A, Tangkijvanich P, Poovorawan Y. Genotypic distribution of hepatitis C virus in Thailand and Southeast Asia. PLoS One May 11, 2015;10(5):e0126764.

[12] Tokita H, Okamoto H, Luengrojanakul P, et al. Hepatitis C virus variants from Thailand classifiable into five novel genotypes in the sixth (6b), seventh (7c, 7d) and ninth (9b, 9c) major genetic groups. J Gen Virol 1995;76:2329–35.

[13] Kamal SM, Nasser IA. Hepatitis C genotype 4: what we know and what we don't yet know. Hepatology April 2008;47(4):1371–83.

[14] Antaki N, Craxi A, Kamal S, Moucari R, Van der Merwe S, Haffar S, Gadano A, Zein N, Lai CL, Pawlotsky JM, Heathcote EJ, Dusheiko G, Marcellin P. The neglected hepatitis C virus genotypes 4, 5 and 6: an international consensus report. Liver Int March 2010;30(3):342–55.

[15] Kamal SM, Mahmoud S, Hafez T, El-Fouly R. Viral hepatitis A to E in South Mediterranean countries. Mediterr J Hematol Infect Dis February 10, 2010;2(1):e2010001.

[16] Jeannel D, Fretz C, Traore Y, et al. Evidence for high genetic diversity and long-term endemicity of hepatitis C virus genotypes 1 and 2 in West Africa. J Med Virol 1998;55:92–7.

[17] Xu LZ, Larzul D, Delaporte E, et al. Hepatitis C virus genotype 4 is highly prevalent in central Africa (Gabon). J Gen Virol 1994;75:2393–8.

[18] Hanafiah M, Groeger K, Flaxman J, Wiersma ST. Global epidemiology of hepatitis C virus infection: new estimates of age-specific antibody to HCV seroprevalence. Hepatology 2013;57:1333–42.

[19] Daw MA, El-Bouzedi A, Ahmed MO, Dau AA, Agnan MM. Hepatitis C virus in North Africa: an emerging threat. Sci World J 2016;2016:7370524.

[20] Ghaderi-Zefrehi H, Gholami-Fesharaki M, Sharafi H, Sadeghi F, Alavian SM. The distribution of hepatitis C virus genotypes in Middle Eastern countries: a systematic review and meta-analysis. Hepat Mon August 23, 2016;16(9):e40357.

[21] Nguyen MH. Prevalence and treatment of hepatitis C virus genotypes 4, 5, and 6. Clin Gastroenterol Hepatol 2005;3(10 Suppl.):S97–101.

[22] Ray SC, Arthur RR, Carella A, et al. Genetic epidemiology of hepatitis C virus throughout Egypt. J Infect Dis 2000;182:698–707.

[23] Pybus OG, Drummond AJ, Nakano T, et al. The epidemiology and iatrogenic transmission of hepatitis C virus in Egypt: a Bayesian coalescent approach. Mol Biol Evol 2003;20:381–7.

[24] Dong ZX, Zhou HJ, Wang JH, Xiang XG, Zhuang Y, Guo SM, Gui HL, Zhao GD, Tang WL, Wang H, Xie Q. Distribution of hepatitis C virus genotypes in Chinese patients with chronic hepatitis C: correlation with patients' characteristics and clinical parameters. J Dig Dis November 2012;13(11):564–70.

[25] Ju W, Yang S, Feng S, Wang Q, Liu S, Xing H, Xie W, Zhu L, Cheng J. Hepatitis C virus genotype and subtype distribution in Chinese chronic hepatitis C patients: nationwide spread of HCV genotypes 3 and 6. Virol J 2015;12:109.

[26] Gu L, Tong W, Yuan M, Lu T, Li C, Lu L. An increased diversity of HCV isolates were characterized among 393 patients with liver disease in China representing six genotypes, 12 subtypes, and two novel genotype 6 variants. J Clin Virol 2013;57:311–7.

[27] Ho SH, Ng KP, Kaur H, Goh KL. Genotype 3 is the predominant hepatitis C genotype in a multi-ethnic Asian population in Malaysia. Hepatobiliary Pancreat Dis Int June 2015;14(3):281–6.

[28] Sievert W, Altraif I, Razavi HA, Abdo A, Ahmed EA, Alomair A, et al. A systematic review of hepatitis C virus epidemiology in Asia, Australia and Egypt. Liver Int 2011;31:61–80.

[29] Kattakuzhy S, Levy R, Rosenthal E, Tang L, Wilson E, Kottilil S. Hepatitis C genotype 3 disease. Hepatol Int November 2016;10(6):861–70.

[30] Valliammai T, Thyagarajan SP, Zuckerman AJ, Harrison TJ. Diversity of genotypes of hepatitis C virus in southern India. J Gen Virol 1995;76:711–6.

[31] Dusheiko G, Schmilovitz-Weiss H, Brown D, McOmish F, Yap PL, Sherlock S, et al. Hepatitis C virus genotypes: an investigation of type-specific differences in geographic origin and disease. Hepatology 2005;19:13–8.

[32] Petruzziello A, Marigliano S, Loquercio G, Cacciapuoti C. Hepatitis C virus (HCV) genotypes distribution: an epidemiological up-date in Europe. Infect Agent Cancer 2016;11:53.

[33] Ruta S, Cernescu C. Injecting drug use: a vector for the introduction of new hepatitis C virus genotypes. World J Gastroenterol October 14, 2015;21(38):10811–23.

[34] Kartasheva V, Döring M, Nietod L, Colettae E, Kaiserf R, Sierraf S, on behalf of the HCV EuResist Study group, et al. New findings in HCV genotype distribution in selected West European, Russian and Israeli regions. J Clin Virol 2016;81:82–9.

[35] Messina JP, Humphreys I, Flaxman A, Brown A, Graham S, Cooke GS, Pybus OG, Barnes E. Global Distribution and prevalence of hepatitis C virus genotypes. Hepatology January 2015;61(1):77–87.

[36] Santos BF, de Santana NO, Franca AV. Prevalence, genotypes and factors associated with HCV infection among prisoners in Northeastern Brazil. World J Gastroenterol July 7, 2011;17(25):3027–34.

Chapter 1.2

Epidemiology and Modes of Transmission of HCV in Developing Countries

Sanaa M. Kamal, Dahlia Ghoraba

Ain Shams Faculty of Medicine, Cairo, Egypt

Chapter Outline

HCV PREVALENCE IN DEVELOPING COUNTRIES

The global prevalence of HCV is 3% with 170 million persons infected worldwide [1–3]. The incidence and prevalence of HCV have significantly declined recently in North America and Western Europe. In the United States, less than 2% of the population is infected [3–5]. In Western Europe, a low prevalence (0.1%) is reported from the United Kingdom, Finland, and Germany and higher prevalence rates >1% are reported from Italy and France [1,2,6]. The prevalence of HCV infection is greater in Eastern Europe (median 2%, range 0.4%–4.9%) [1,3,7,8].

High prevalence rates are reported from many developing countries (Table 1.2.1). South Asia has high rates of HCV infection [9]. In China, the estimated HCV prevalence ranges between 0.29% and 9.6% according to the region [10–13] (Fig. 1.2.1). The incidence has been estimated at 24.2/100,000 [13]. In Taiwan, HCV prevalence ranges between 2.9% and 17.0% [14,15]. A study by Lee et al. [15] reported a prevalence of 2.6%–30.9% in some areas. In hyperdynamic areas in Taiwan, the incidence rates of HCV infection were 110 per 10,000 persons. There is also evidence of high endemicity (3.8%–7.5%) among tribes in Northern Thailand [16,17].

Hepatitis C in Developing Countries. https://doi.org/10.1016/B978-0-12-803233-6.00002-3

TABLE 1.2.1 Seroprevalance of Hepatitis C in Different Developing Countries

Country	%	Country	%
Algeria	2.0	Mexico	0.7
Angola	**3.9**	Morocco	1.6
Argentina	0.6	Malawi	2,0
Brazil	1.3	Mali	1.9
Burkina Faso	**6.1**	Mozambique	1.3
Burundi	3.1	Namibia	1.6
Cameroon	**4.9**	Nigeria	3.1
Chile	0.9	**Pakistan**	**5.9**
China	2.2	Philippines	3.6
DR of the Congo	2.9	Romania	4,5
Egypt	**14.7**	Rwanda	3.1
Ethiopia	2.7	Senegal	1.0
Gabon	**4.9**	Sudan	3.2
Gambia	2.4	Somalia	2.6
Ghana	3.2	South Africa	1.1
Georgia	**6.7**	Tanzania	2.7
Guinea	1.5	Thailand	2.2
India	1.5	Tunisia	1.8
Indonesia	2.1	Turkey	0.9
Iran	0.2	Uganda	2.7
Ivory Coast	2.2	Ukraine	1.2
Kenya	2.8	**Uzbekistan**	**6.5**
Libya	1.2	Zambia	1.1
Madagascar	1.7	Zimbabwe	1.6

The prevalence among injecting drug users (IDU) was 86%–95% [18]. The blood donor study reported a prevalence of 1.37% and comprised voluntary donors from five separate studies [19,20]. In Vietnam, HCV prevalence was estimated at 2.0%–2.9% among adults. HCV prevalence of 0.8% was reported

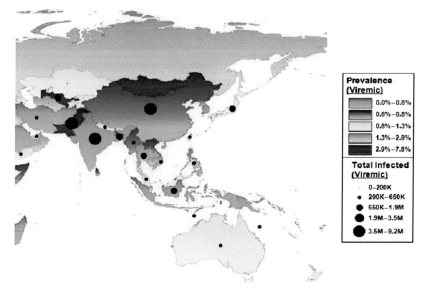

FIGURE 1.2.1   Hepatitis C in Central and South Asia.

among blood donors in Hanoi and 21% among blood donors in Ho Chi Minh City [21,22], suggesting an increasing prevalence from North to South.

In India, the estimated HCV prevalence ranges between 1% and 1.9% [23,24] with significant variation between regions. The northern part of the country had similar practices and risk factors as Pakistan, where prevalence was above 2% [24]. Among IDUs in Northern India, prevalence was reported to be 33.7% [24]. Hemodialysis patients and thalassemics receiving multiple blood transfusions showed a prevalence of 41.9% and 25.45%, respectively [25]. Patients with sexually transmitted diseases had a prevalence of 2.6% [25,26]. In Pakistan, a meta-analysis reported prevalence of 3% among blood donors and 4.7% in the general population [27]. Another study estimated an overall prevalence of 3% in adults, 2.8% in adult blood donors, 5.4% in adult nonblood donors, and 2.1% in children [28]. HCV positivity rates as high as 30% were found in Punjab, the largest Pakistani province [29,30] (Fig. 1.2.2).

The magnitude of the problem of HCV in Sub-Saharan Africa is not well documented. However, there is strong evidence that HCV reaches high levels and represents a heavy burden and public health threat in low-income regions. Estimates suggest that sub-Saharan Africa accounts for nearly 20% of global infections [31–33]. A meta-analysis that included more than 10 years' data showed that the prevalence of HCV was 7.8%, 4.5%,3% and 0.8% in Central Africa, West Africa, East Africa and South Africa respectively. The countries with the highest prevalence include Cameroon (13.8%), Burundi (11.3%), Gabon (9.2%), Uganda (6.6%), Guinea (5.5%), and Ghana (3%) [31,33,34] (Fig. 1.2.3). Blood donors consistently had the

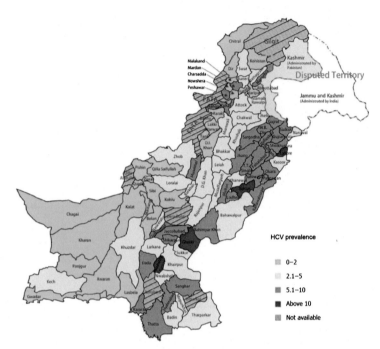

FIGURE 1.2.2 Hepatitis C in Pakistan.

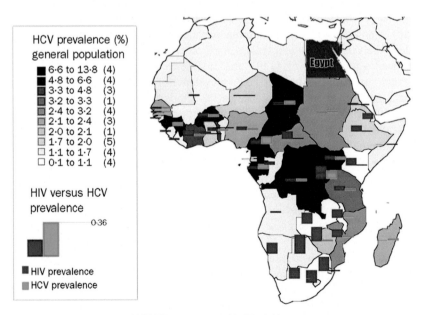

FIGURE 1.2.3 Hepatitis C in Africa.

lowest prevalence (1.78%), followed by pregnant women (2.51%), individuals with comorbid HIV (3.57%), individuals from the general population (5.41%), those with a chronic illness (7.99%), and those at high risk for infection (10.18%) [34,35].

Egypt is the country with the highest HCV prevalence in the world. According to a large national survey conducted in 2015, the prevalence of HCV was 10%, which is lower than the prevalence of 14.5% reported in the 2008 nationwide survey. The overall prevalence was 17.4% and 12.2% in males and females, respectively, and increases with age [36,37]. The Egyptian Health Ministry estimated the HCV incidence to be 6.9/1000 persons per year based on regression modeling [38]. HCV accounts for 31% of acute viral hepatitis cases in Egypt [39]. Very high rates of infection are clustered in the Nile Valley despite the active preventive procedures, better blood screening measures, and better sanitization practices within hospitals. The large pool of infected individuals is the reservoir that maintains continued transmission [40].

AGE AND GENDER DISTRIBUTION IN DEVELOPING COUNTRIES

Specific data on the age, gender, racial, and ethnic distributions of HCV infection in developing countries are scarce. In developing countries in Asia and Africa, HCV is usually newly acquired at the second or third decade [14,15]. With the tendency of HCV toward chronic evolution, the prevalence of chronic HCV increases with age, with the highest rate being reported in the age group older than 40 years. In Egypt, high rates of infection are observed in all age groups including young individuals, indicating an ongoing high risk for acquiring HCV infection [16–18]. More than 60% of acute HCV infections are in persons below the age of 25 years. High incidence rates (14.1 per 1000 person-years) have been detected in Egyptian children younger than 10 years of age living in households with an anti-HCV–positive parent [19,20]. This high incidence in young persons could lead to future increases in chronic disease in these individuals and persistence of the high magnitude of the burden of HCV-related chronic disease in Egypt. The prevalence of acute HCV in both genders is controversial. Although some studies showed higher HCV incidence among men [21,22] other population-based surveys [23,24] showed slightly higher rates in women than in men. However, additional epidemiologic studies are needed to confirm that the risk of HCV transmission is greater in men.

MODES OF TRANSMISSION: CHANGING PATTERNS IN THE WEST AND EAST

Several modes of transmission of HCV are well documented and widely accepted; others are less well defined and require further study. HCV is most frequently transmitted through direct percutaneous exposure to infected blood. Risk factors for HCV transmission differ between developed and developing countries. Socioeconomic differences are likely to explain much of the geographic variability.

Blood Transfusion

Percutaneous inoculation via transfusion is very efficient in transmitting HCV infection. Transfusion-associated cases occurred before routine donor screening in blood banks with second- and third-generation enzyme immunoassays, a procedure that resulted in a sharp decline in transfusion-associated HCV transmission [25]. As of 2001, the risk of HCV infection from a unit of transfused blood is less than one per million transfused units. However, the risk of HCV transmission through blood has not been fully eliminated in some developing countries that lack the resources to implement adequate donor screening and continue to use commercial donors to supplement their blood supplies [26]. In such countries, blood transfusion still accounts for some percentage of acute HCV cases.

Injecting Drug Users

The number of cases of acute hepatitis C among injection drug users has recently declined in the United States and Europe because of the widespread implementation of harm reduction policies, syringe exchange programs, counseling of IDUs regarding protection from infection, and changing injecting behavior [27,28]. However, both incidence and prevalence of HCV infection remain high among drug injectors in developing and transitional countries with high IDU prevalence such as Eastern Europe and Central and South Asia [29–31]. Lower prevalence rates of IDU are reported from the Middle East where IDU does not represent the major risk factor for acute HCV. Engagement in unsafe injection practices such as syringe sharing, sharing of injection equipment, engaging in sexual relationships with other IDUs increase the risk of HCV transmission also accounts for a significant percentage of acute HCV cases [39] (Fig. 1.2.4).

Sexual Transmission

Sexual transmission of HCV has been controversial. Some case–control studies identified some well-documented instances of acute hepatitis C occurring after a defined sexual exposure [41,42]. Studies showed that HCV sexual transmission increases with the number of lifetime sexual partners, high-risk sexual exposure, and unprotected sex [43,46]. The prevalence of HIV in several resource-limited countries in Asia and Africa has been shown to be related to HCV prevalence in such countries [43,44].

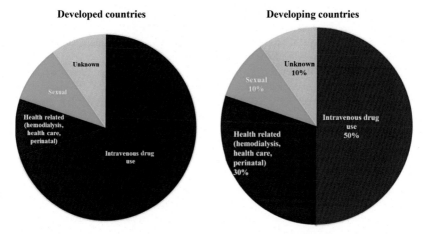

FIGURE 1.2.4 Hepatitis C modes of transmission in developing and developed countries.

In countries with high HCV endemicity, relatively higher rates of sexual transmission have been reported and reflect the higher background prevalence in such countries. In rural Egypt, sexual transmission between monogamous spouses [41] range between 3% and 34% (95% CI: 0% to 49%). Acute HCV is associated with a high temporal risk of transmission of HCV to sexual partners. In some studies, sexual transmission was confirmed in studies by phylogenetic analysis in 15% of sexual partners of individuals with acute HCV [45,46]. However, although some cases of acute HCV have been related to sexual transmission, the degree to which sexual transmission of HCV occurs is controversial because sexual transmission is difficult to confirm given that partners might have other risk factors for HCV transmission such as IDU. Phylogenetic analysis to identify genetic relatedness of HCV viral isolates in partners is required to confirm sexual transmission. Studies on the sexual transmission of hepatitis C are limited by the potential of the confounding variable of IDU or shared items such as razors and other items among sexual partners. Large studies directed at evaluating HCV-infected persons with multiple sexual partners are needed for accurate estimates of sexual transmission.

REFERENCES

[1] WHO. WHO-Hepatitis C. Geneva, Switzerland: World Health Organization; 2015.
[2] Mohd Hanafiah K, Groeger J, Flaxman AD, Wiersma ST. Global epidemiology of hepatitis C virus infection: new estimates of age-specific antibody to HCV seroprevalence. Hepatology 2013;57:1333.
[3] Centers for Disease Control and Prevention. Surveillance for viral hepatitis. 2015. https://www.cdc.gov/hepatitis/statistics/2015surveillance/commentary.htm.
[4] Alter MJ. Epidemiology of hepatitis C. Hepatology 1997;26:62S.

[5] Armstrong GL, Wasley A, Simard EP, et al. The prevalence of hepatitis C virus infection in the United States, 1999 through 2002. Ann Intern Med 2006;144:705.

[6] Cornberg M, Razavi HA, Alberti A, et al. A systematic review of hepatitis C virus epidemiology in Europe, Canada and Israel. Liver Int 2011;31(Suppl. 2):30–60.

[7] Blachier M, Leleu H, Peck-Radosavljevic M, et al. The burden of liver disease in Europe: a review of available epidemiological data. J Hepatol 2013;58:593–608.

[8] Madaliński K, Zakrzewska K, Kołakowska A, Godzik P. Epidemiology of HCV infection in Central and Eastern Europe. Przegl Epidemiol 2015;69(3):459–64. 581–4.

[9] Wait S, Kell E, Hamid S, Muljono DH, Sollano J, Mohamed R, Shah S, Mamun-Al-Mahtab, Abbas Z, Johnston J, Tanwandee T, Wallace J. Hepatitis B and hepatitis C in southeast and southern Asia: challenges for governments. Lancet Gastroenterol Hepatol November 2016;1(3):248–55.

[10] Qin Q, Smith MK, Wang L, Su Y, Wang L, Guo W, Wang L, Cui Y, Wang N. Hepatitis C virus infection in China: an emerging public health issue. J Viral Hepat 2015;22(3):238–44.

[11] Cui Y, Jia J. Update on epidemiology of hepatitis B and C in China. J Gastroenterol Hepatol August 2013;28(Suppl. 1):7–10.

[12] Zhang Y. Li-Min Chen, Miao He Hepatitis C Virus in mainland China with an emphasis on genotype and subtype distribution. Virol J 2017;14:41.

[13] Cai LN, Zhu SW, Zhou C, Wang YB, Jiang NZ, Chen H, Tang XY, Wang JH, Chen X, Hu WJ, et al. Infection status of HBV, HCV and HIV in voluntary blood donors of Chinese Nanjing area during 2010–2013. Zhongguo Shi Yan Xue Ye Xue Za Zhi 2014;22:1089–93.

[14] Bennett H, Waser N, Johnston K, Kao JH, Lim YS, Duan ZP, Lee YJ, Wei L, Chen CJ, Sievert W, et al. A review of the burden of hepatitis C virus infection in China, Japan, South Korea and Taiwan. Hepatol Int 2015;9:378–90.

[15] Lee C, Nan Lu S, Hung CH, Tung WC, et al. Hepatitis C virus genotypes in southern Taiwan: prevalence and clinical implications. Trans R Soc Trop Med Hyg 2006;100:767–74.

[16] Wasitthankasem R, Posuwan N, Vichaiwattana P, Theamboonlers A. Decreasing hepatitis C virus infection in Thailand in the past decade: evidence from the 2014 national survey. PLoS One 2016;11(2):e0149362.

[17] Chimparlee N, Oota S, Phikulsod S, Tangkijvanich P, Poovorawan Y. Hepatitis B and hepatitis C virus in Thai blood donors. Southeast Asian J Trop Med Public Health 2011;42: 609–15.

[18] Ratanasuwan W, Sonji A, Tiengrim S, Techasathit W, Suwanagool S. Serological survey of viral hepatitis A, B, and C at Thai Central Region and Bangkok: a population base study. Southeast Asian J Trop Med Public Health 2004;35:416–20.

[19] Sunanchaikarn S, Theamboonlers A, Chongsrisawat V, Yoocharoen P, Tharmaphornpilas P, Warinsathien P, et al. Seroepidemiology and genotypes of hepatitis C virus in Thailand. Asian Pac J Allergy Immunol 2007;25:175–82.

[20] Akkarathamrongsin S, Hacharoen P, Tangkijvanich P, Theamboonlers A, Tanaka Y, Mizokami M, et al. Molecular epidemiology and genetic history of hepatitis C virus subtype 3a infection in Thailand. Intervirology 2013;56:284–94.

[21] Dunford L, Carr M, Dean J, Waters A, et al. Hepatitis C virus in Vietnam: high prevalence of infection in dialysis and multi-transfused patients involving diverse and novel virus variants. PLoS One 2012;7(8):e41266.

[22] Berto A, Day J, Chau NV, Thwaites GE, et al. Current challenges and possible solutions to improve access to care and treatment for hepatitis C infection in Vietnam: a systematic review. BMC Infect Dis 2017;17:260.

[23] Puri P, Anand A, Saraswat VA, Acharya SK, et al. Consensus Statement of HCV Task Force of the Indian National Association for Study of the Liver (INASL). Part I: status report of HCV infection in India. J Clin Exp Hepatol June 2014;4(2):106–16.

[24] Satsangi S, Chawla YK. Viral hepatitis: Indian scenario. Med J Armed Forces India 2016; 72(3):204–10.

[25] Gupta V, Kumar A, Sharma P, Bansal N, et al. Most patients of hepatitis C virus infection in India present late for interferon-based antiviral treatment: An epidemiological study of 777 patients from a North Indian tertiary care center. J Clin Exp Hepatol June 2015;5(2):134–41.

[26] Kumar A, Acharya SK, Singh SP. The Indian National Association for study of the liver (INASL) consensus on prevention, diagnosis and management of hepatocellular carcinoma in India: the recommendations. J Clin Exp Hepatol 2014;4(Suppl. 3):S3–26.

[27] Umer M, Iqbal M. Hepatitis C virus prevalence and genotype distribution in Pakistan: comprehensive review of recent data. World J Gastroenterol 2016;22(4):1684–700.

[28] Waheed Y, Shafi T, Safi SZ, Qadri I. Hepatitis C virus in Pakistan: a systematic review of prevalence, genotypes and risk factors. World J Gastroenterol 2009;15:5647–53.

[29] Umar M, Bushra HT, Ahmad M, Data A, Ahmad M, Khurram M, Usman S, Arif M, Adam T, Minhas Z, et al. Hepatitis C in Pakistan: a review of available data. Hepat Mon 2010;10: 205–14.

[30] Ahmed F, Irving WL, Anwar M, Myles P, Neal KR. Prevalence and risk factors for hepatitis C virus infection in Kech District, Balochistan, Pakistan: most infections remain unexplained. A cross-sectional study. Epidemiol Infect 2012;140:716–23.

[31] Karoney MJ, Siika AM. Hepatitis C virus (HCV) infection in Africa: a review. Pan Afr Med J 2013;14:44.

[32] Madhava V, Burgess C, Drucker E. Epidemiology of chronic hepatitis C virus infection in sub-Saharan Africa. Lancet Infect Dis May 2002;2(5):293–302.

[33] Agyeman AA, Akousa A, Mprah A, Ashiagbor A. Epidemiology of hepatitis C virus in Ghana: a systematic review and meta-analysis. BMC Infect Dis 2016;16:391.

[34] Riou J, Aït Ahmed M, Blake A, Vozlinsky S, Brichler S, Eholié S, Boëlle PY, Fontanet A. HCV epidemiology in Africa group. Hepatitis C virus seroprevalence in adults in Africa: a systematic review and meta-analysis. J Viral Hepat 2016;23(4):244–55.

[35] Rosnay B, Galani T, Njouom R, Moundipa PF. Hepatitis C in Cameroon: what is the progress from 2001 to 2016? J Transl Int Med December 1, 2016;4(4):162–9.

[36] Gomaa A, Allam N, Elsharkway A, El Kassas M, Waked I. Hepatitis C infection in Egypt: prevalence, impact and management strategies. Hepat Med 2017;9:17–25.

[37] Mohamoud Y, Mumtaz G, Riome S, Miller D, Raddad L. The epidemiology of hepatitis C virus in Egypt: a systematic review and data synthesis. BMC Infect Dis 2013;13:288.

[38] Ministry of Health and Population [Egypt], El-Zanaty and Associates [Egypt], ICF International. Egypt health issues survey 2015. 2015. Cairo: Ministry of Health and Population; Rockville, MD: ICF International.

[39] Kamal SM. Acute hepatitis C: a systematic review. Am J Gastroenterol 2008;103(5):1283–97.

[40] El-Zanaty F, Way A. Egypt demographic and health survey 2008. Cairo: Ministry of Health, El-Zanaty and Associates and Macro International; 2009. p. 431.

[41] Midgard H, Weir A, Palmateer N, Lo Re 3rd V, Pineda JA, Macías J, Dalgard O. HCV epidemiology in high-risk groups and the risk of reinfection. J Hepatol October 2016;65(1 Suppl.):S33–45.

[42] Chan DP, Sun HY, Wong HT, Lee SS. Hung CC Sexually acquired hepatitis C virus infection: a review. Int J Infect Dis August 2016;49:47–58.

[43] Alter MJ. HCV routes of transmission: what goes around comes around. Semin Liver Dis 2011;31(4):340–6.

[44] Mboto CI, Davies A, Fielder M, Jewell AP. Human immunodeficiency virus and hepatitis C co-infection in sub-Saharan West Africa. Br J Biomed Sci 2006;63(1):29 37.

[45] Miller FD, Elzalabany MS, Hassani S, Cuadros DF. Epidemiology of hepatitis C virus exposure in Egypt: opportunities for prevention and evaluation. World J Hepatol December 8, 2015;7(28):2849–58.

[46] Kamal SM, Amin A, Madwar M, Graham CS, He Q, Al Tawil A, Rasenack J, Nakano T, Robertson B, Ismail A, Koziel MJ. Cellular immune responses in seronegative sexual contacts of acute hepatitis C patients. J Virol November 2004;78(22):12252–8.

Section 2

The burden of Hepatitis C in Developing countries

Public Health and Economic Burden of Hepatitis C Infection in Developing Countries

Tamer A. Hafez
American University, Cairo, Egypt

Chapter Outline

INTRODUCTION

Globally, an estimated 3% of the world's population equating to 170–180 million persons are living with hepatitis C virus (HCV) infection [1,2]. The characteristic feature of HCV infection is its tendency to persist, resulting in chronic hepatitis that progresses in some individuals to cirrhosis, end-stage liver disease, or hepatocellular carcinoma (HCC) [3]. Approximately, 70%–80% of acute HCV-infected individuals will develop chronic liver disease and, within 10–30 years, approximately 20% of them will experience cirrhosis and its complications, of which 1%–5% are HCC each year [3,4]. Chronic hepatitis C and its sequels represent major causes of liver-related morbidity and mortality particularly in countries with a high burden of infection.

Most of the hepatitis C cases occur in middle income and low income countries [5; Fig. 2.1.1] and the majority of related deaths occur in resource-limited,

Hepatitis C in Developing Countries. https://doi.org/10.1016/B978-0-12-803233-6.00003-5

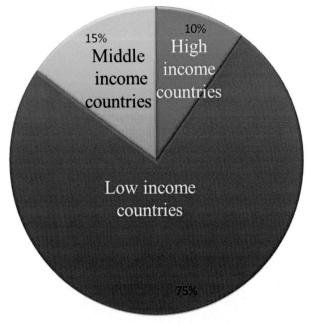

FIGURE 2.1.1 Distribution of people living with Hepatitis C in low, middle, and high income countries.

underprivileged regions in Africa and Asia [5]. Although HCC is the fifth most common malignancy worldwide, it is considered the most common cancer in countries with heavy burden of HCV such as Egypt and several countries in Africa [6,7]. In developing countries, HCV-related HCC develops in young individuals mostly between the ages of 30 and 40 years [6,7].

Disease burden is the impact of a health problem as measured by financial cost, mortality, morbidity, or other indicators. It is often quantified in terms of quality-adjusted life years or disability-adjusted life years (DALYs), both of which quantify the number of years lost because of disease (YLDs) [8]. The burden of disease and the death rate associated with HCV are expected to double over the next 2 decades as this large cohort of HCV-positive patients progress to cirrhosis, cancer, and liver failure. In resource-limited countries, HCV infection, chronic hepatitis, cirrhosis, and HCC have serious economic consequences that include the direct health-care expenditures in caring for the disease and indirect costs in terms of loss of productivity or absence from work because of chronic hepatitis C infection, cirrhosis, and HCC [9,10]. Although the rate of new HCV infection in the United States and Europe declined dramatically, the rates are still high in many developing countries [11]. In developing and resource-limited countries in Africa, the Middle East, South Asia, and Latin America, the prevalence of HCV still exceeds 2% [12,13]. Egypt has the highest HCV prevalence rates, which exceed 10% [12–19]. Despite the growing evidence that hepatitis C is an urgent public health issue, few countries have developed strategic national responses to address the hepatitis C epidemics within their populations.

Impact of the Modes of Transmission on the Burden of HCV in Developing Countries

The modes of HCV transmission differ between developed and developing countries. In developed countries, injection drug use and sharing drug injection equipment are the leading *means* of *HCV transmission* [20]. Whereas in developing countries, HCV transmission frequently results from exposure to infected blood and blood products in the health-care and community settings [21,22]. Unsafe injections and poor infection control in hospitals result in 2 million new HCV infections each year [22]. HCV prevalence reaches 60%–80% in high-risk populations such as drug users and HIV-positive individuals [23].

In some developing countries, specific traditional practices such as circumcision, home deliveries, and scarification contribute to HCV transmission. In Asia, the virus is mostly transmitted through the use of injecting drugs, whereas in Africa, epidemiologic data suggest an important transmission through unsafe medical procedures and injections [21,22].

Factors Maintaining Heavy Burden of HCV in Developing Countries

The persistence of transmission and maintenance of HCV in several developing countries with high incidence and prevalence of HCV infection are attributed to several factors that include deficient prevention and screening strategies, inadequate health education, lack of strict infection control in health facilities, difficult assessment of the disease progression, in addition to insufficient access to treatment at affordable prices [21].

Coinfections play an important role in increasing the burden of HCV in developing and resource-limited countries. It is estimated that up to 30% of the 33 million persons infected with HIV also are infected with HCV [22]. The prevalence of HIV–HCV coinfection is significantly high in sub-Saharan Africa and several Asian countries. Coinfection with HIV is associated with accelerated progression of liver disease and increased mortality among HCV-infected persons [23,24]. Some studies suggest that HCV/HIV coinfection may increase the risk of mother-to-child HCV transmission. In Egypt, HCV and schistosomiasis coinfection is associated with accelerated fibrosis and cirrhosis, higher mortality, higher rates of HCC, and lower quality of life [18,25,26].

Economic Burden of HCV in Egypt

Limited studies assessed the economic burden of HCV in Egypt. Estes et al. [26], constructed a model to assess the direct health care cost and indirect burden of HCV in Egypt between 2013 and 2030. The direct health-care cost was calculated based on data collected from the National Liver Institute, which is a representative of governmental hospitals. Indirect costs were estimated based on calculating DALYs, which include YLD and years of life lost because of premature death; in addition, the value of lost productivity was estimated using Egyptian estimates for the value of a statistical life year (VSLY). The same scenarios previously used in Waked et al. (2014) to estimate the economic burden were used by Estes et al. [26] to calculate the economic burden as well, where the first scenario had the old treatment with 48% sustained virologic response (SVR) and the number of cases treated annually is about 65,000; the second scenario is with 90% SVR and 65,000 cases treated annually; and the third scenario had 90% SVR and 325,000 cases treated annually.

Other Types of Disease Burden

Lack of awareness about hepatitis C in the community often leads to misinformation, missing of opportunities for prevention and treatment, and stigmatization of infected populations. The level of support that someone with chronic hepatitis C might receive was less when compared with someone with a chronic illness that does not carry a stigma which might affect self-esteem and cause a decrease in the quality of life. This stigma of the disease pushes 30.9% of the patients to conceal the nature of their illness and 24.4% of them to even conceal this information from their families, according to a study conducted in Elghar village, Zagazig governorate. Also, patients with HCV are not allowed to travel to Gulf countries for work. This opportunity lost because of the inability of traveling for working in certain countries is not included in our analysis.

Impact of Barriers to Treatment Access on the Burden of HCV in Developing Countries

Viral eradication has been demonstrated to significantly decrease the risk of cirrhosis and its complications, including HCC. A study estimated that the risk of liver-related death could be reduced by 36% if all infected patients were treated. However, there is lack of consensus about HCV screening and treatment guidelines in developing countries with high prevalence of HCV. Furthermore, the limited infrastructure hinders proper HCV assessment and treatment. The limited access to treatment and the delay in diagnosis significantly increase treatment costs because many patients seek medical care at advanced stages of the disease. Early stage of the disease is less expensive for treatment but later stages and their complications are associated with higher treatment costs. In patients with chronic hepatitis C infection, nearly 70% of direct costs are allocated to medical diagnostic services (such as laboratory tests, taking biopsy, ultrasonography, etc.), followed by the cost of drugs (16%). However, hospital admissions for patients in the later stages of the disease (cirrhosis and HCC) were among the highest direct costs. This can be attributed to the duration of admissions and stay in the intensive care units (ICUs). A study found that the average medical expenditures associated with having an HCV diagnosis peak early after the first-observed diagnosis, but that diagnosis may come relatively late in the disease, leading to high costs. More specifically, inpatient and outpatient costs were over USD 10,000 per year in the first 2 years following the diagnosis.

Patients in several resource-limited countries do not have comprehensive medical insurance that covers HCV treatment. Because of the high financial burden of the disease, many patients ignore their disease for a long time, resulting in disease progression that increases the requirement for hospital admissions. Therefore, direct and indirect costs rose with increased length of hospital stay. This is because costs of ICUs are higher than other wards. A group of patients who discover their disease at earlier stages are obliged to borrow from relatives, sell their assets, or get a loan to get treatment. Thus, HCV therapy imposes a heavy toll on their household income.

Pegylated interferon (IFN) therapy–ribavirin (RBV) combined therapy has been the standard of care for HCV-infected patients for decades. However, Peg-IFN–RBV therapy has been associated with serious adverse events, noncompliance, and low SVR in genotypes 1 and 4. Clinical care for patients with HCV has advanced remarkably during the last 2 decades, as a result of better understanding of the pathophysiology of the disease, developments in diagnostic procedures, and improvements in therapy and prevention. After the introduction of direct-acting antiviral agents (DAAs), the treatment paradigm radically changed and very high sustained virologic rates were achieved. Unlike IFN regimens, which rely on upregulating the patients own immune system, these DAAs block different stages of viral replication.

Economic Challenges Associated With DAAs

Despite the excellent potency of DAAs, their very high cost represents a serious barrier for the use of DAAs in most resource-limited settings and may prove an obstacle to eradication of HCV infection worldwide. Logically, successful treatment initiated at the early stages of chronic hepatitis C should prevent many late HCV complications, which incur extremely high costs. However, because the prevalence of HCV is high in most resource-limited countries, rationing of HCV treatment is crucial. Patients at high risk of disease transmission, progression, and development of complications need to be prioritized. Targeting populations with high HCV prevalence like drug users, prisoners, and migrants also makes sense because they are most likely to spread infection to others. But those most likely to transmit to others often have low disease stage. Besides, they are most likely to abrogate the personal benefits of treatment by being reinfected. Treating those who have the most symptoms also makes sense, but unfortunately symptoms do not always improve with cure. However, prioritization may have negative repercussion in the communities of developing countries because every patient requests management of his disease.

Thus, it is critical to make DAAs cost-effective and accessible to all eligible patients. Of countries with heavy HCV burden, Egypt and India succeeded in producing generic versions of the DAAs sofosbuvir and daclatasvir, which are being mass produced for <1% of the current US retail price. This initiative made some DAAs available for a higher proportion of patients. Other countries such as China conducted several clinical trials to shorten the duration of therapy. A study in China demonstrated 100% SVR with triple DAAs for only 3 weeks if the patient without cirrhosis achieved ultra-rapid virologic response (HCV RNA < 500 IU 48 h after starting treatment).

Primary Prevention Campaigns and the Burden of HCV

In the absence of efficient vaccines and the prohibiting costs of recent HCV therapies, primary prevention of HCV infection, particularly in the health-care setting, represents a very important approach for reducing the burden of HCV infection in developing countries. Inadequate control of HCV transmission in such countries will result in continuous rise in HCV-associated morbidity and mortality from cirrhosis and HCC for years or decades.

Comprehensive prevention and control programs that address HCV transmission should be the first priority for any country. Nationwide screening and preventive campaigns should integrate testing, care, and treatment of HCV infection and should be carefully tailored to meet the needs of individual countries.

In conclusion, HCV infection continues to a major public health and economic burden in developing countries. Reducing the burden of HCV-related disease can be achieved through a comprehensive approach that incorporates primary prevention of transmission through enhanced infection control and injection safety in health-care settings and in the community, universal screening of blood and blood products, harm reduction programs, and increased public awareness about risk factors for HCV infection. For the already infected persons, newer, more effective therapies are available. However, lack of access to screening, care, and treatment limits the use of these therapies for most persons living with HCV infection globally, and deaths from preventable cirrhosis and liver cancer continue to increase. Governments need to address viral hepatitis comprehensively by improving surveillance, prevention, care, and treatment.

ABBREVIATIONS

DAAs Direct-acting antiviral agents
HCC Hepatocellular carcinoma
HCV Hepatitis C virus
PEG-IFN Pegylated interferon
RBV Ribavirin

REFERENCES

[1] Lavanchy D. Evolving epidemiology of hepatitis C virus. Clin Microbiol Infect 2011;17:107–15.

[2] GBD 2013 Mortality and Causes of Death Collaborators. Global, regional, and national age-sex specific all-cause and cause-specific mortality for 240 causes of death, 1990–2013: a systematic analysis for the Global Burden of Disease Study 2013. Lancet 2015;385(9963):117–71.

[3] Seeff LB. The history of the "natural history" of hepatitis C (1968–2009). Liver Int January 2009;29(Suppl. 1):89–99.

[4] Thomas DL, Seeff LB. The natural history of hepatitis C. Clin Liver Dis August 2005;9(3):383–98.

[5] Lemoine M, Thursz M. Hepatitis C, a global issue: access to care and new therapeutic and preventive approaches in resource-constrained areas. Semin Liver Dis February 2014;34(1):89–97.

[6] Jaka H, Mshana SE, Rambau PF, Masalu N, Chalya PL, Kalluvya SE. World J Surg Oncol August 2, 2014;12:246.

[7] Sartorius K, Sartorius B, Aldous C, Govender PS, Madiba TE. Global and country underestimation of hepatocellular carcinoma (HCC) in 2012 and its implications. Cancer Epidemiol June 2015;39(3):284–9.

[8] WHO. Global burden of disease [Internet] Available at: http://www.who.int/topics/global_burden_of_disease/en/.

[9] Younossi ZM, Kanwal F, Saab S, Brown KA, El-Serag HB, Kim WR, Ahmed A, Kugelmas M, Gordon SC. The impact of hepatitis C burden: an evidence-based approach. Aliment Pharmacol Ther March 2014;39(5):518–31.

[10] Wedemeyer H, Dore GJ, Ward JW. Estimates on HCV disease burden worldwide – filling the gaps. J Viral Hepat January 2015;22(Suppl. 1):1–5.

[11] Messina JP, Humphreys I, Flaxman A, Brown A, Cooke GS, Pybus OG, Barnes E. Global distribution and prevalence of hepatitis C virus genotypes. Hepatology January 2015;61(1):77–87.

[12] Kershenobich D, Razavi HA, Sanchez-Avila JF, et al. Trends and projections of hepatitis C virus epidemiology in Latin America. Liver Int 2011;31(Suppl. 2):18–29.

[13] Kamal SM. Improving outcome in patients with hepatitis C virus genotype 4. Am J Gastroenterol 2007;102(11):2582–8.

[14] Miller FD, Abu-Raddad LJ. Evidence of intense ongoing endemic transmission of hepatitis C virus in Egypt. Proc Natl Acad Sci 2010;107(33):14757–62.

[15] Ministry of Health and Population [Egypt], El-Zanaty and Associates [Egypt], ICF International. Egypt health issues survey 2015. 2015.

[16] Doss W, Mohamed MK, Esmat G, El Sayed M, Fontanet A, Cooper S, El Sayed N. Egyptian national control strategy for viral hepatitis 2008–2012. Arab Republic of Egypt, Ministry of Health and Population, National Committee for the Control of Viral Hepatitis; 2014.

[17] Alter MJ. Epidemiology of hepatitis C virus infection. World J Gastroenterol 2007;13:2436–41.

[18] Thursz M, Fontanet A. HCV transmission in industrialized countries and resource-constrained areas. Nat Rev Gastroenterol Hepatol January 2014;11(1):28–35.

[19] Gower E, Estes C, Blach S, et al. Global epidemiology and genotype distribution of the hepatitis C virus infection. J Hepatol 2014;61:S45–57.

[20] Platt L, Easterbrook P, Gower E, et al. Prevalence and burden of HCV co-infection in people living with HIV: a global systematic review and meta-analysis. Lancet Infect Dis 2016.

[21] Lacombe K, Rockstroh J. HIV and viral hepatitis coinfections: advances and challenges. Gut 2012;61(Suppl. 1):i47–58.

[22] Bruggmann P, Berg T, Øvrehus AL, Moreno C, Brandão Mello CE, Roudot-Thoraval F, Marinho RT, Sherman M, Ryder SD, Sperl J, Akarca U, Balık I, Bihl F, Bilodeau M, Blasco AJ, Buti M, Calinas F, Calleja JL, Cheinquer H, Christensen PB, Clausen M, Coelho HS, Cornberg M, Cramp ME, Dore GJ, Doss W, Duberg AS, El-Sayed MH, Ergör G, Esmat G, Estes C, Falconer K, Félix J, Ferraz ML, Ferreira PR, Frankova S, García-Samaniego J, Gerstoft J, Giria JA, Gonçales Jr FL, Gower E, Gschwantler M, Guimarães Pessôa M, Hézode C, Hofer H, Husa P, Idilman R, Kåberg M, Kaita KD, Kautz A, Kaymakoglu S, Krajden M, Krarup H, Laleman W, Lavanchy D, Lázaro P, Marotta P, Mauss S, Mendes Correa MC, Müllhaupt B, Myers RP, Negro F, Nemecek V, Örmeci N, Parkes J, Peltekian KM, Ramji A, Razavi H, Reis N, Roberts SK, Rosenberg WM, Sarmento-Castro R, Sarrazin C, Semela D, Shiha GE, Sievert W, Stärkel P, Stauber RE, Thompson AJ, Urbanek P, van Thiel I, Van Vlierberghe H, Vandijck D, Vogel W, Waked I, Wedemeyer H, Weis N, Wiegand J, Yosry A, Zekry A, Van Damme P, Aleman S, Hindman SJ. Historical epidemiology of hepatitis C virus (HCV) in selected countries. J Viral Hepat May 2014;21(Suppl. 1):5–33.

[23] Sievert W, Altraif I, Razavi HA, Abdo A, Ahmed EA, Alomair A, Amarapurkar D, Chen CH, Dou X, El Khayat H, Elshazly M, Esmat G, Guan R, Han KH, Koike K, Largen A, McCaughan G, Mogawer S, Monis A, Nawaz A, Piratvisuth T, Sanai FM, Sharara AI, Sibbel S, Sood A, Suh DJ, Wallace C, Young K, Negro F. A systematic review of hepatitis C virus epidemiology in Asia, Australia and Egypt. Liver Int July 2011;31(Suppl. 2):61–80.

[24] El-Shabrawi MH, Kamal NM. Burden of pediatric hepatitis C. World J Gastroenterol November 28, 2013;19(44):7880–8.

[25] Ansaldi F, Orsi A, Sticchi L, Bruzzone B, Icardi G. Hepatitis C virus in the new era: perspectives in epidemiology, prevention, diagnostics and predictors of response to therapy. World J Gastroenterol August 7, 2014;20(29):9633–5.

[26] Estes C, Abdel-Kareem M, Abdel-Razek W, Abdel-Sameea E, Abuzeid M, Gomaa A, Osman W, Razavi H, Zaghla H, Waked I. Economic burden of hepatitis C in Egypt: the future impact of highly effective therapies. Aliment Pharmacol Ther September 2015;42(6):696–706.

Chapter 2.2

Social, Cultural, and Political Factors Influencing HCV in Developing Countries

Sara A. Abdelhakam[1], Mohamad A. Othman[2]

[1]Ain Shams University, Cairo, Egypt; [2]University of Gezira, Khartoum, Sudan

Chapter Outline

INTRODUCTION

Several sociocultural issues are associated with the different aspects and phases of hepatitis C virus (HCV) infection. The vast majority of HCV research is focused on epidemiology, natural history, clinical aspects, and treatment of hepatitis C. Qualitative research investigating the impact of social and cultural factors on the quality of life of patients living with HCV is scarce. Clinical experience with patients, observations, and ongoing community-based *participatory* and ethnographic research demonstrated that stigmatization of those infected with HCV is a complex social phenomenon involving individual and societal judgments. Patients infected with HCV face a number of social and psychological challenges, such as living with a contagious potentially life-threatening illness and tolerating stigmatization and discrimination in their family and community [1,2]. In developing and developed countries, stigma usually associates people infected with HCV with illicit drug use, criminality, promiscuity, untrustworthiness, and HIV [3,4]. This stigma is so pervasive that individuals who accidentally acquire the infection through iatrogenic transmission or blood

Hepatitis C in Developing Countries. http://dx.doi.org/10.1016/B978-0-12-803233-6.00004-7
33

transfusion do not disclose their HCV status to others in their social networks and may not seek treatment [5–7].

INFLUENCE OF SOCIAL FACTORS ON HCV TRANSMISSION IN DEVELOPING COUNTRIES

In Egypt, the country with the highest HCV prevalence in the world, the hepatitis C epidemic is an explicit example of iatrogenesis when medical interventions could harm patients. The epidemic was initiated and propagated by the extensive antischistosomiasis mass treatment campaigns undertaken in 1918 and 1982 to treat and eradicate the endemic parasitic infection schistosomiasis [9]. Millions of Egyptians infected by schistosomiasis were treated with consecutive injections of potassium antimony tartrate administered as a series of intravenous injections using improperly sterilized syringes and needles that were reused [9–11]. Although the mass antischistosomal campaign was terminated in the early 1980s, the prevalence and incidence of HCV remains high in Egypt.

Blood transfusion was a common route for HCV transmission in developed and developing countries and was responsible for considerable numbers of HCV infection before obligatory screening of blood and blood products for HCV [11a,11b]. However, unsafe blood transfusion remains a risk in some developing countries. Occupational transmission in health care workers through needlesticks and sharps injuries is common and contributes to the high rates of HCV infections among health care providers in Egypt, Pakistan, and Nigeria [12–14a].

Another set of risk factors mostly related to prevailing social, cultural, and religious conservatism contribute to the high prevalence of HCV in several developing countries. Women particularly in rural and semiurban communities likely acquire HCV infection through health-related practices to which they are exposed as a result of deeply rooted traditions. Prevailing ultraconservative concepts and high levels of illiteracy among rural women further increase the gender gap, gender inequality, and ill-health among women infected with HCV. In rural Egypt, Pakistan, Sudan, and India, about 50% of deliveries are attended to by traditional midwives or birth assistants who do not perform appropriate cleaning or disinfection of the equipment [15]. Reluctance of women to deliver in health facilities is largely attributed to social, cultural, and religious reasons because in several semiurban and rural communities women and families do not feel comfortable to be attended to by male gynecologists and prefer to be assisted by local midwives.

Male circumcision is widely practiced in health facilities in Muslim countries shortly after birth. However, officially female circumcision is banned by law but the majority of families in rural areas and Upper Egypt still practice it as they have a deep conviction that female circumcision is a religious and moral obligation. Sometimes, circumcision of several girls is conducted during special celebrations with family gathering. Most families that circumcise their daughters seek the services of traditional informal providers or midwives to avoid being penalized. The midwives do not follow hygienic measures or infection control procedures during the process, endangering the lives of the young girls and subjecting them not only to acquiring a variety of blood-borne infections but also to the risk of bleeding to death or shock. A survey showed that despite banning female circumcision, 76% of the women surveyed exposed their daughters to circumcision [16,17]. Adequate training and continuous education of registered midwives on infection control measures may be a reasonable compromise between the people's traditions and safety measures.

LAY PERSPECTIVE OF HCV IN DEVELOPING COUNTRIES

In several developing and low-resource countries, patients feel that they are victims of medical procedures, but at the same time, they seek biomedical therapies to achieve cure. Medicalization of most aspects of life and the medical dominance of physicians in addition to the power of the pharmaceutical companies and media created a state of psychological dependence on prescription drugs or professional medical opinion [18,21]. As a result, patients feel powerless and unable to handle their own health anymore because of the inherent conviction that they are "futureless forgotten victims." Freidson [19] called for greater recognition of patient or lay views in health issues and showed the fundamental differences between the medical and social approaches in understanding illness. Therefore, some mutual understanding should be forged between the lay beliefs and expert knowledge to maintain a physician–patient dialogue.

The lay perspectives and ideologies play an important role in shaping the response and reaction of individuals to HCV infection. Poverty plays an important role in the response of HCV patients to their disease. Some patients feel that they are helpless and cannot afford seeking decent medical care. The accidental discovery of the disease is extremely frustrating and devastating for an apparently healthy individual. In many instances, an individual first knows that he/she is infected with HCV during an employment medical examination. The underlying social beliefs, concepts, and fears related to disease progression, transmission to family members, and inequalities in accessing health care completely change the lives, performance, assumptions, and beliefs of infected individuals. Adding to the disappointment and frustration is the high numbers of patients for whom no source of infection is apparent.

A peculiar physician–patient relationship is observed in patients with hepatitis C in developing countries. Physicians in most cases do not spend time with their patients discussing the nature of the disease, risk of transmission to the family and community, the natural history, disease progression patterns, treatment options, alternatives, or expectations. Some physicians might even blame patients for their illness because the patients do not follow the health precautions provided by professionals or because they do not get regularly tested. In turn, patients blame physicians for being arrogant and uncaring [20].

HEPATITIS C AND INJECTING DRUG USE: THE REALITIES OF STIGMATIZATION AND DISCRIMINATION

In several developing countries, particularly Arab and Islamic countries, illicit drug use is considered by the society and law as a criminal, immoral, anti-religious, and shameful conduct that should be penalized and prosecuted. According to the conservative values of such communities, the social discourse blames drug users because they "brought the disease upon themselves" and do not deserve care or compassion [8]. Injecting drug users (IDUs) are also blamed for acquiring and placing "innocent people" at risk of fatal infections such as HIV and hepatitis [8]. The stigma associated with injecting drug use makes it difficult for the IDUs to talk about their problems even with their families. Thus, IDUs experience continuous denial and extreme vigilance to prevent discovery of their addiction and infection problems. Ultimately, fear of stigmatization may cause the drug users to avoid being tested and those with chronic HCV to avoid treatment. IDUs represent real health threats given that they are not recognized and are away from the preventive, surveillance, and treatment programs. The unrestricted mobility of IDUs within the community in addition to the continuous change of injecting locations and patterns increase their capability of transmitting HCV infection.

DISCRIMINATION TOWARD PEOPLE LIVING WITH HEPATITIS C

Several developing countries impose hepatitis B and C testing before employment, and infected persons are not recruited. The Gulf countries, Kingdom of Saudi Arabia, Kuwait, Qatar, and Emirates prohibit recruiting workers positive for HCV. Pakistanis, Indians, and Egyptians seek work in the oil-rich countries. Ironically, many of those applying for work in these countries knew by chance that they are infected during preemployment screening. As a result, patients with hepatitis C who are apparently healthy and capable to work are disqualified from employment. They feel discriminated, victimized, and isolated. The accidental discovery of the infection coupled with possible stigma and inability to take appropriate job subject the patients to severe psychological trauma that makes them entirely hopeless, angry, or extremely anxious to be treated but could not access therapy because of inequalities.

In conclusion, HCV infection in developing countries is a multifaceted public health, social, and economic problem. The low socioeconomic conditions, poor public health programs, misconceptions, high levels of illiteracy, and deeply rooted traditional and cultural beliefs contribute to the increased burden of HCV in several developing countries. Stigma poses significant challenges to patients infected with hepatitis C, their families, and community. Discrimination, inequalities in accessing health services, and expensive recent antiviral agents increase the burden of HCV in developing countries. Understanding the social determinants of hepatitis C is necessary for understanding not only the modes of transmission but also the impact of the disease on the patients' behaviors and treatment decisions.

REFERENCES

[1] Paterson B, Backmund M, Hirsch G, Yim C. The depiction of stigmatization in research about hepatitis C. Int J Drug Policy 2007;18:364–73.

[2] Fraser S, Treloar C. Spoiled identity' in hepatitis C infection: the binary logic of despair. Crit Public Health 2006;16:99–110.

[3] Treloar C, Rhodes T. The lived experience of hepatitis C and its treatment among injecting drug users: qualitative synthesis. Qual Health Res 2009;19:1321–34.

[4] Sladden TJ, Hickey AR, Dunn TM, Beard JR. Hepatitis C virus infection: impacts on behaviour and lifestyle. Aust New Zeal J Public Health 1998;22(4):509–11.

[5] Dunne EA, Quayle E. Pattern and process in disclosure of health status by women with iatrogenically acquired hepatitis C.. J Health Psychol 2002;7(5):565–82.

[6] Crofts N, Louie R, Loff B. The next plague: stigmatization and discrimination related to hepatitis C virus infection in Australia. Health Hum Rights 1997;2:87–96.

[7] Herek GM, Capitanio JP, Widaman KF. Stigma, social risk, and health policy: public attitudes toward HIV surveillance policies and the social construction of illness. Health Psychol 2003;22(5):533–40.

[8] Zickmund S, Ho E, Masuda M, Ippolito L, LaBrecque D. "They treated me like a leper": stigmatization and the quality of life of patients with hepatitis C. J Gen Intern Med 2003;18:835–44.

[9] Frank C, Mohamed MK, Strickland GT, Lavanchy D, Arthur RR, Magder LS, El Khoby T, Abdel-Wahab Y, Anwar W, Sallam I. The role of parenteral antischistosomal therapy in the spread of hepatitis C virus in Egypt. Lancet March 11, 2000;355(9207):887–91.

[10] El-Sayed HF, Abaza SM, Mehanna S. The prevalence of hepatitis B and C infections among immigrants to a newly reclaimed area endemic for Schistosoma mansoni in Sinai, Egypt. Acta Trop 1997;68:229–37.

[11] Hyams KC, Mansour MM, Massoud A, Dunn MA. Parenteral antischistosomal therapy: a risk factor for hepatitis B infection. J Med Virol 1987;23:109–14.

[11a] Grebely J, Dore G. What is killing people with hepatitis C virus infection? Semin Liver Dis 2012;31:331–9.

[11b] Kandeel AM, Talaat M, Afifi SA, El-Sayed NM, Abdel Fadeel MA, Hajjeh RA, Mahoney FJ. Case control study to identify risk factors for acute hepatitis C virus infection in Egypt. BMC Infect Dis November 12, 2012;12:294.

[12] Onyekwere CA, Hameed L. Hepatitis B and C virus prevalence and association with demographics: report of population screening in Nigeria. Trop Doct October 2015;45(4):231–5.

[13] Stoszek SK, Abdel-Hamid M, Narooz S, El Daly M, Saleh DA, Mikhail N, et al. Prevalence of and risk factors for hepatitis C in rural pregnant Egyptian women. Trans R Soc Trop Med Hyg 2006;100:102–7.

[14] Ezechi OC, Kalejaiye OO, Gab-Okafor CV, Oladele DA, Oke BO, Musa ZA, Ekama SO, Ohwodo H, Agahowa E, Gbajabiamilla T, Ezeobi PM, Okwuraiwe A, Audu RRA, Okoye RN, David AN, Odunkwe NN, Onwujekwe DI, Ujah IA. Sero-prevalence and factors associated with Hepatitis B and C co-infection in pregnant Nigerian women living with HIV infection. Pan Afr Med J March 13, 2014;17:197.

[14a] Qureshi H. Prevalence of hepatitis B and C viral infections in Pakistan: findings of a national survey appealing for effective prevention and control measures. East Mediterr Health J 2010;16:15.

[15] Murad EA, Babiker SM, Gasim GI, Rayis DA, Adam I Epidemiology of hepatitis B and hepatitis C virus infections in pregnant women in Sana'a, Yemen. BMC Pregnancy Childbirth June 7, 2013;13:127.

[16] Assad M. Female circumcision in Egypt. Stud Fam Plan January 1980;11(1):3–6.

[17] Baasher T. Psycho-social aspects of female circumcision. In: Baasher T, Bannerman R, Rushwan H, Sharaf I, editors. Traditional health practices affecting the health of women and children. vol. 2 (2). WHO Regional Office for the Eastern Mediterranean. EHO EMRO Technical Publication; 1982. p. 162–80.

[18] Engel GL. The need for a new medical model: a challenge for biomedicine. Science 1997;196(4286):129–36.

[19] Freidson E. Profession of medicine: a study of the sociology of applied knowledge. Chicago (IL): University of Chicago Press; 1997.

[20] Zickmund S, Hillis SL, Barnett MJ, Ippolito L, LaBrecque DR. Hepatitis C virus infected patients report communication problems with physicians. Hepatology 2004;39(4):999–1007.

[21] Maturo A. Medicalization: current concept and future directions in a bionic society. Mens Sana Monogr 2012 January–December;10(1):122–33.

Section 3

Hepatitis C Infection in Specific Geographic Regions

Chapter 3.1

Hepatitis C in Egypt

Sanaa M. Kamal[1], Sara A. Abdelhakam[2]

[1]Ain Shams Faculty of Medicine, Cairo, Egypt; [2]Ain Shams University, Cairo, Egypt

Chapter Outline

INTRODUCTION

Hepatitis C virus (HCV) infection is a major health and economic problem in Egypt. Large studies have been conducted on the various aspects of HCV in Egypt in the 2 decades since HCV diagnostic assays became available. Several studies showed very high prevalence of HCV antibodies ranging from 10% to 20% [5,8–10]. The high prevalence of HCV infection is attributed in part to parenteral antischistosomal treatment campaigns that were conducted until the 1970s. However, after stoppage of such campaigns, epidemiologic studies suggest that HCV continues to be transmitted at high rates [1–3]. Prospective cohort studies monitoring HCV seroconversion among susceptible persons have identified incidence rates ranging from 3.1 to 5.2 per 100,000 [4,6,7,11,12]. Based on these studies, it is projected that 248,000–416,000 infections are expected to emerge each year in Egypt.

Epidemiologic studies demonstrated diverse exposures associated with HCV infection in Egypt, including unsafe injections, health-care procedures, and various community exposures. Given the heavy burden of HCV in Egypt, the Egyptian government and Egyptian Ministry of Health (EMOH) launched huge nationwide campaigns to treat patients with chronic HCV with the latest and most effective therapies.

PREVALENCE AND INCIDENCE OF HCV IN EGYPT

Prevalence

Egypt has the highest prevalence of hepatitis HCV in the world, with almost 10 million Egyptians exposed to the virus and about 5–7 million active infections. The major cause of the HCV epidemic in Egypt was the extensive nationwide schistosomiasis control programs that used intravenous tartar emetic [8]. The HCV epidemic in Egypt was discovered through HCV screening of 2164 blood donors using second-generation enzyme-linked immunosorbent assay, which first became available in Egypt in 1992. In this screening, HCV prevalence was 10.9% were HCV antibody positive. The reported prevalence was almost 10 times the prevalence in other parts of the world [9]. A rural population–based community study that screened the entire population of a small rural village in the northern Nile Delta revealed an overall HCV antibody prevalence of 17.6% [10]. Another large cross-sectional study [5] investigated the prevalence of antibodies to HCV in a rural Egyptian community in the Nile Delta. Blood samples were obtained from 3,888 (75.4%) of 5156 residents. Overall, 973 (24.3%) of 3999 residents were anti-HCV–positive, and the age- and gender-adjusted seroprevalence was 23.7%. Anti-HCV prevalence increased sharply with age, from 9.3% in those 20 years of age and younger to >50% in those older than 35 years. Currently or previously married individuals were more likely to be seropositive than those never married, controlling for age [Mantel–Haenszel risk ratio = 1.8; 95% confidence interval (CI): 1.3, 2.6]. Of the 905 anti-HCV–positive samples tested, 65% were also positive for HCV RNA. Active schistosomal infection was not associated with anti-HCV status; however, history of antischistosomal injection therapy (reported by 19% of anti-HCV positives) was a risk for anti-HCV (age-adjusted risk ratio = 1.3; 95% CI: 1.2, 1.5). Thus, many studies [1–6,9,10] were conducted all over Egypt, all of which showed an alarmingly high prevalence of HCV throughout Egypt with varying provenances in the country (Fig. 3.1.1).

The EMOH conducted several large national surveys investigating the prevalence of HCV among the general population in Egypt [6,7]. In the 2008 Egypt Demographic and Health Survey, 4757 households including 12,780 individuals aged 15–59 years were screened for HCV. The HCV antibody prevalence was 14.7% and chronic hepatitis with positive HCV RNA polymerase chain reaction test was detected in 9.8% of screened individuals [7].

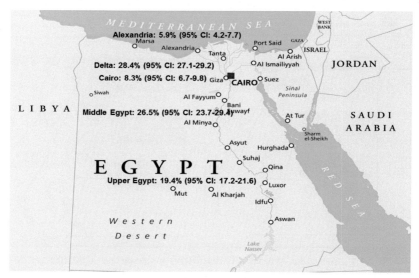

FIGURE 3.1.1 The prevalence of hepatitis C virus in different geographic regions in Egypt before 2008.

In 2015, the Egypt Health Issues Survey (EHIS) screened 7649 households including 28,079 individuals. Screening demonstrated prevalence of anti-HCV and HCV RNA in 6.3% and 4.4%, respectively [11]. The prevalence of HCV among males was more than that among females, and the prevalence of active hepatitis showed a positive relationship with age. A significant difference in HCV prevalence rates was observed between the older age groups and younger age group. The prevalence of active hepatitis in those aged below 20 was less than 1%, whereas it reached 22.1% among those aged 55–59 years [11].

Comparing HCV prevalence rates reported in the 2008 and 2015 surveys showed an overall significant reduction of 32% and 29% in the prevalence of HCV antibody and HCV RNA-positive individuals, respectively, between the DHS in 2008 and the EHIS in 2015 (Fig. 3.1.2). A statistically significant reduction in HCV antibody prevalence was observed in all age groups, and the greatest relative prevalence reduction (75%) was observed among those aged 15–19 years. A statistically significant reduction in HCV RNA–positive cases was noted (Figs. 3.1.2 and 3.1.3).

When comparing the results of both surveys, a significant reduction in the overall prevalence of HCV antibody from 14.7% to 10.0%, and HCV RNA from 9.9% to 7.0%, was observed between 2008 and 2015 among those aged 15–59 years. It was also found that the residence, socioeconomic level, and education standards have a significant impact on the prevalence of HCV. According

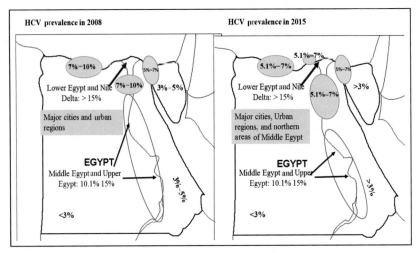

FIGURE 3.1.2   Comparison of hepatitis C virus (HCV) prevalence rates in the 2008 and 2015 national surveys.

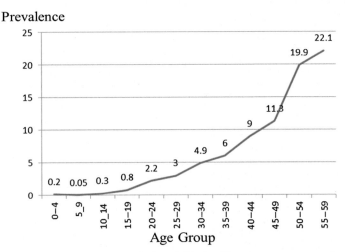

FIGURE 3.1.3   Prevalence of hepatitis C virus (HCV) in different age groups according to the 2015 survey.

to the 2015 survey, the prevalence of active hepatitis cases in rural areas was higher than urban areas, where it is 5.1% and 3.1%; respectively. In addition, the prevalence in lower Egypt governorates is the highest (5.6%), while urban governorates showed the lowest prevalence (3%). Also, it was clear from the results that the prevalence of chronic hepatitis C cases increases in patients with limited financial resources [11].

This reduction in prevalence between 2008 and 2015 may be attributed to several factors such as the marked reduction in numbers of the old patients initially infected during the mass schistosomiasis treatment campaign with reused syringes (1960s through early 1980s) to outside the age range covered by the survey (i.e., those older than 59 years) [11–14]. The successful nationwide public health campaigns contributed to increasing the awareness of Egyptians for avowing the risk factors. The ambitious large treatment program also played an important role in reducing HCV in Egypt.

PREVALENCE OF HCV AMONG HIGH-RISK GROUPS IN EGYPT

The prevalence of HCV significantly increases in the high-risk groups in Egypt. These groups include patients with schistosomiasis, patients on hemodialysis, patients with multitransfusion, patients with thalassemia, injecting drug users (IDUs), health-care workers, and barbers. Treatment of patients with schistosomiasis by using parenteral antischistosomal therapy (PAT) campaigns is believed to be the major historic cause for increase in the prevalence of HCV in Egypt [12]. The average HCV prevalence among patients with schistosomiasis treated with PAT 20 years ago ranged between 38% and 84.0% [1–3]. More recent studies [15] reported HCV prevalence of 7.7% among patients with schistosomiasis who were treated orally up to 8 years ago. Several studies [16–22] were conducted to identify the prevalence of HCV among children with multitransfusion. High HCV prevalence rates were observed with averages of about 42% among children with multi-transfusion and about 58% among children with thalassemia [16–22]. The prevalence of HCV is very high in patients undergoing hemodialysis with prevalence rates ranging between 70% and 90% depending on the Egyptian regions [23–25]. The burden of disease among health-care workers is also high, where HCV prevalence was about 17% in average [26–28]. In rural Egypt, barbers perform some health-related procedures such as drainage of abscesses and circumcision. A study reported HCV prevalence of 12.3% for rural barbers [29]. HCV status has not been adequately studied in IDUs. Few studies estimated that almost 63.0% of IDUs were HCV positive [30–32] (Fig. 3.1.4).

Incidence

The incidence of HCV in Egypt varies according to the geographic region. Studies conducted in villages in Qalubyia, Lower Egypt, Assuit, and Upper Egypt reported high HCV incidence rates of 6.8/1000 person-years (PY) and 0.8/1000 PY [1]. Another 4-year cohort study was conducted in Menoufia governorate, Lower Egypt, where the incidence was 2.4/1000 PY (95% CI: 1.6, 3.5) [33]. Studies [12,34] estimated the incidence in specific groups such as children and pregnant women. A total of 2852 uninfected infants from three villages with high HCV prevalence were prospectively followed from

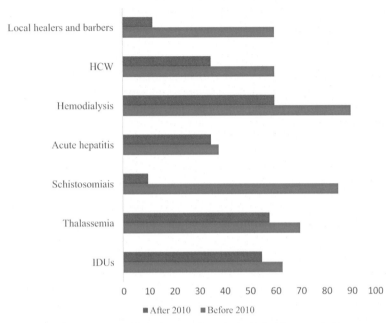

FIGURE 3.1.4   Hepatitis C virus prevalence in the high-risk groups in Egypt.

birth for up to 5.5 years, and the incidence rate was found to be 2.7/1000 PY. The incidence of HCV among pregnant women was 5.2/1000 PY [33].

Some studies attempted to establish mathematical models to estimate the incidence of HCV in Egypt. In 2010, Miller et al. [34] estimated that HCV incidence was 6.9/1000 (95% CI: 5.5, 7.4) PY equating to more than 500,000 new cases of HCV. On the other hand, in 2013, another model was conducted to address the assumption of time-independent epidemiology to reach a more realistic estimate. This model estimated the incidence of HCV nationwide at around 2.0/1000 PY, which means around 150,000 new infections annually [34].

In 2014, a mathematical model was proposed to estimate the present and future burden of HCV [12]. A disease progression model was developed using Microsoft Excel where the model started with new acute hepatitis C cases that develop chronicity in addition to chronic viremic HCV case. These cases go through the model and progress from one stage to another according to certain probabilities as in Fig. III5. It moves between different stages of fibrosis (F0, F1, F2, and F3) till it reaches the cirrhotic stage, then it develops decompensated cirrhosis, which has different forms such as variceal hemorrhage, hepatic encephalopathy, and ascites.

To date, it is not clear if a reduction of HCV incidence contributed to the decline in HCV prevalence observed in some Egyptian regions. As shown in 2008 and 2015 [7,11], the prevalence of HCV antibodies in those aged 15–19 years was 4.1% and 1.0%, respectively. The percentage of relative risk reduction was 75% (95% CI: 64, 85), implying a very substantial reduction in HCV incidence in the past 20 years in this age group. Furthermore, the prevalence of HCV antibody (0.4%) and HCV RNA-positive persons (0.2%) observed in 2015 among those aged 1–14 years is low compared with several studies conducted in early 2000s, which described the HCV antibody prevalence to be ranging from 2% to 7% in children younger than 10 years in rural areas.

This change in incidence in the younger age groups could possibly be because of the various public health interventions implemented by the Ministry of Health and Population and its partners [14]. Several efforts have been made to promote and expand the infection prevention and control programs beyond Ministry of Health and Population hospitals and university hospitals. Auto-disabled syringes were introduced to the routine immunization sector in 2008 to promote safe injection practices among children. Safe blood transfusion activities, including policies and guidelines, have been intensified. Raising the awareness of the public, by targeting universities and schools to improve their understanding on the epidemiology and prevention of viral hepatitis, was also carried out. Educational programs targeting health-care staff has been carried out to enforce the concepts of safe health care and prevention of blood-borne pathogens [14].

ACUTE HEPATITIS C IN EGYPT

Acute HCV infection is typically defined as a new occurrence of viremia with conversion from an HCV RNA–negative to an HCV-positive status. Acute HCV infection is asymptomatic in the majority of cases [35–40]. HCV RNA can be detected in the serum in almost all patients within 1–2 wk of exposure. Seroconversion is detected after 2–6 months (window period) or later in certain risk groups, making anti-HCV testing less reliable than HCV RNA assessments for early diagnosis of acute HCV [20–24]. The acute phase of infection is usually considered to be the first 6 months, and this is the phase of infection in which spontaneous clearance is still possible. Although 20% to 50% of patients with acute hepatitis achieve spontaneous resolution, the majority of acute HCV cases evolve to chronic disease [35–41].

Acute hepatitis C represents 30% of the overall acute viral hepatitis cases reported in Egypt [42]. More than 60% of acute HCV infections are in persons younger than of 25 years. High incidence rates (14.1 per 1000 PY) have been detected in Egyptian children younger than 10 years of age living in households with an anti-HCV–positive parent [42,43]. This high incidence in young persons could lead to future increases in chronic disease in these individuals and persistence of the high magnitude of the burden of HCV-related chronic disease in Egypt. The prevalence of acute HCV in both genders is controversial. Some studies showed higher HCV incidence among men [44]; however, this was not further confirmed in other studies.

MODES OF TRANSMISSION OF HCV IN EGYPT

Several modes of transmission of HCV are well documented in Egypt and others are less well defined. Percutaneous inoculation via blood transfusion has sharply declined since the obligatory implementation of blood screening in blood banks. HCV transmission through IDUs is still considered as a mode of HCV transmission.

Sexual transmission of HCV has been controversial. In Egypt, relatively higher rates of sexual transmission have been reported and reflect the higher background prevalence in this country. In rural Egypt, sexual transmission between monogamous spouses [44] ranged between 3% and 34% (95% CI: 0%, 49%). Acute HCV is associated with a high temporal risk of transmission of HCV to sexual partners. Sexual transmission was confirmed by phylogenetic analysis in 15% of sexual partners of individuals with acute HCV [45,46].

Occupational Exposures

Infection with HCV is an important occupational hazard for health-care workers. The risk of occupational HCV transmission increases with deep injuries and after procedures involving hollow-bore needles [47,48]. In longitudinal studies, attempts have been made to assess the risk of infection associated with occupational exposures to HCV in the health-care setting. In these studies, transmission rates ranged from 0% to 10.3% [28,47–49]. The substantial variation in transmission rates in Egypt is based on differences in implementation of infection control policies and occupational infection surveillance systems. In Egypt, occupational transmission among health-care workers through needlesticks and sharp injuries contributes to new HCV cases, given that needlestick prevention devices are not yet adopted in all the hospitals and health-care units in addition to inadequate compliance with universal, standard precautions in some health facilities [28].

In Egypt, intrafamilial HCV transmission is considered an important route of transmission where living in a house with an infected family member is a risk factor for HCV transmission. Analysis of data collected during surveys of Egyptian rural communities show that children whose parents had antibodies to HCV were at higher risk for contracting anti-HCV than children whose parents did not [44] (Fig. 3.1.5).

From a public health standpoint, health-care systems should adopt strategies to identify individuals at high risk for HCV infection and provide opportunities

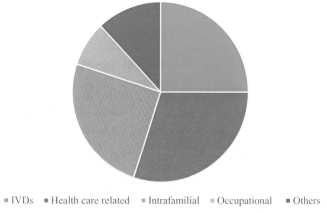

■ IVDs ■ Health care related ■ Intrafamilial ■ Occupational ■ Others

FIGURE 3.1.5 Modes of hepatitis C virus (HCV) transmission in Egypt.

for education and behavior modification. In Egypt, surveillance programs and efforts to increase awareness improve diagnosis and facilitate treatment of acute HCV, which will have far-reaching implications for the management of chronic HCV, where current disease management is very expensive.

IMPACT OF IMMIGRATION ON HCV TRANSMISSION

North Africa is currently an active point for immigration toward Europe. Although North Africa including Egypt is not a main source of illegal immigrants, this region is considered an important transit region used by many Africans illegally moving to European countries as their final destiny. Recent studies conducted to estimate the prevalence of HCV together with HIV and hepatitis B virus (HBV) on African immigrants showed that the overall prevalence of HCV in the African immigrants was 7.6%, though it varied according to the origin of the immigrants, ranging from 5.7% to 10.0% [50]. The prevalence of HCV in Central African immigrants was 5.7%, ranging from 3.6% to 8.2%. Of the samples retrieved from WA immigrants, 8.1% were positive for HCV antibody, ranging from 2.1% to 14.1%. The prevalence was high in the subjects from Nigeria and Ghana (14.2%), followed by Burkina Faso/Ivory Coast (8.3%) and Niger/Mali (7.4%). The prevalence in immigrants from HOA was (8.4%), with a range of 6.8%–9.9%; it was 6.3%, 8.2%, and 9.3% for subjects from Eritrea, Somalia, and Ethiopia, respectively. In NA immigrants, the prevalence was 10.0% and ranged from 1.3% to 18.7%. The Egyptian immigrants and Nigerians showed the highest prevalence rates, while the Maghreb immigrants and Chadians showed the lowest prevalence rates [51,52]. Thus, the African influx either as transient immigrants or as residents represents a threat for the preventive efforts taken by the Egyptian government.

ECONOMIC BURDEN OF HCV IN EGYPT

Limited studies assessed the economic burden of HCV in Egypt. Estes et al. [27] constructed a model to assess the direct health-care cost and indirect burden of HCV in Egypt between 2013 and 2030. The direct health-care cost was calculated based on data collected from the National Liver Institute, which is a representative of governmental hospitals. Indirect costs were estimated based on calculating disability-adjusted life years, which includes years of life lost to disability and years of life lost because of premature death; in addition, the value of lost productivity was estimated using Egyptian estimates for the value of a statistical life year. The scenarios used previously in Waked et al. 2014 [28] to estimate the economic burden were used by Estes et al. (2015) to calculate the economic burden as well, where the first scenario had the old treatment with 48% sustained virologic response (SVR) and treated cases annually is about 65,000; the second scenario was considered with 90% SVR and 65,000 treated cases annually; while the third scenario had 90% SVR and 325,000 treated cases annually (Table 3.1.1).

TABLE 3.1.1 Direct and Indirect Costs in Egypt, 2015–30

Cumulative disability-adjusted life years 2013–30 % of change from scenario (1)	7,875,440	7,343,640 −6.8%	4,923,210 −37.5%
Cumulative direct costs 2013–2030 (US $) % of change from scenario (1)	23,244,377,860	24,192,586,440 4.1%	18,632,607,710 −19.8%
Cumulative indirect costs 2013–2030 (US $) % of change from scenario (1)	65,822,552,110	61,547,348,360 −6.5%	38,929,874,750 −40.9%
Cumulative total costs 2013–2030 (US $) % of change from scenario (1)	89,066,929,970	85,739,934,800 −3.7%	57,562,482,460 −35.4%

Adapted with permission from Estes C, Abdel-Kareem M, Abdel-Razek W, Abdel-Sameea E, Abuzeid M, Gomaa A, Osman W, Razavi H, Zaghla H, Waked I. Economic burden of hepatitis C in Egypt: the future impact of highly effective therapies. Aliment Pharmacol Ther September 2015;42(6):696–706.

TREATMENT OF HEPATITIS C IN EGYPT IN THE ERA OF DIRECT-ACTING ANTIVIRAL AGENTS

Egypt has the highest burden of HCV infection worldwide. According to the 2008 national survey, about 15% of the population were seropositive, 10% chronically infected, with 90% of patients infected with genotype 4 [7]. Given the serious complications of chronic hepatitis C and the huge burden imposed by HCV-related advanced liver disease, the Egyptian government adopted an enthusiastic campaign since 2008 for combating the HCV epidemic. This campaign included treating established infection, detection of recent infections, and preventing new infections. In 2015, the seroprevalence of HCV infection in Egypt has declined to 6.3% among the studied populations with an overall estimated 30% decrease in HCV prevalence in Egypt between 2008 and 2015 [11]. However, HCV transmission is still ongoing, and incidence rates have been estimated at 2.4 per 1000 PY (close to 165,000 new infections annually). In some reports, prevalence rates were estimated at 6.9/1000 PY based on a regression model using a national probability sample [6].

Initially, the standard of care used in the national treatment campaign included a combination of pegylated interferon (PEG-IFN) and ribavirin (RBV). However, the SVR did not exceed 60% of patients of genotype 4. The noncompliance, several adverse effects of PEG-IFN and RBV in addition to the various contraindications for HCV therapy contributed to the suboptimal performance of PEG-IFN and RBV. The emergence of antiviral agents represented a breakthrough in the control of HCV in Egypt. The high efficacy of DAAs and the all-oral regimen represented the hope for combating HCV in this hardly hit country. The Egyptian government and the National Committee for Control of Viral Hepatitis exerted extensive efforts for making the most effective medications available at affordable costs for treating large numbers of HCV patients.

The initial step in implementing the nationwide HCV treatment program included establishment of a Web-based online registration system Web site (www.nccvh.org.eg) once the first DAA was registered in Egypt. The system has an appropriate bandwidth to accommodate extensive data entry and applications. The portal was designed for registration of patients with HCV and scheduling appointments at the treatment centers all over Egypt. The patient enters his/her name, national ID, and residence. Patients' appointments were automatically set to the first time available in the treatment center nearest to their residence. Scheduled lists were seen online at the treatment centers and included patients' national ID data. In the first week of the Web-based system launch, more than 300,000 were registered. By mid-2016, there is no wait time for the first appointment in any of the Egyptian centers, and patients are evaluated on the same week of registration all over Egypt.

When the patient visits the treatment center, he or she is subjected to a thorough history taking, physical examination, abdominal ultrasound, transient elastography, and a panel of laboratory investigations (Fig. 3.1.6).

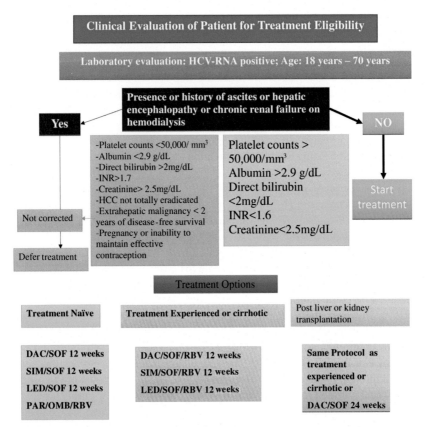

FIGURE 3.1.6 Algorithm for Patients' evaluation and management at the Egyptian hepatitis C virus (HCV) treatment center. *DAC,* daclatasvir; *LED,* ledipasvir; *OMB,* ombitasvir; *PAR,* paritaprevir; *RBV,* ribavirin; *SIM,* simeprevir; *SOF,* sofosbuvir.

Patients eligible to antiviral therapy are informed about the treatment plan, duration of treatment, adverse events, and the follow-up schedule. All treated patients are instructed to visit the HCV medical centers on a monthly basis for follow-up. DAA combinations are prescribed according to the patients' status. Treatment-naive patients are treated with 12 weeks of one of the following regimen: Daclatasvir and sofosbuvir (SOF) or simeprevir (SIM) with SOF or ledipasvir plus SOF or paritaprevir plus ombitasvir plus RBV.

Treatment-experienced patients who failed previous therapy or patients with liver cirrhosis are treated with any of the previous regimen with the addition of RBV. Patients with chronic hepatitis C after liver or kidney transplantation are treated with the same protocol followed for treating patients with cirrhosis, but for 24 weeks. Patients are required to return after completion of the course of treatment for evaluation of SVR according to the treatment protocol.

At the end of treatment, patients are advised about the prevention of HCV reinfection and the importance of HBV vaccination. Patients with cirrhosis are regularly monitored following treatment for detection of complications of cirrhosis or HCC, and they have liver tests, ultrasonography, and tests for estimating the alpha fetoprotein levels carried out every 6 months.

HEPATITIS C PREVENTION STRATEGIES IN EGYPT

The EMOH realized the importance of coupling the HCV treatment efforts with halting the ongoing transmissions of HCV. An action plan addressing the different aspects of viral hepatitis control was launched. The action plan targeted all aspects of viral hepatitis prevention and control: disease surveillance; infection control and injection safety; blood safety; vaccination; information, education, and communication; screening, care, and treatment; research; and governance. The plan updated guidelines for infection control and blood safety and implemented training programs in infection control and adoption of strategies for injection safety and blood safety.

In conclusion, HCV has been a major public health and economic problem in Egypt. However, the enormous efforts of the Egyptian government in combating the HCV epidemic are considered a successful disease control program serving millions of patients. This model provides access to all HCV-infected individuals and an action plan and strategy for prevention of new infections.

ABBREVIATIONS

DAAs Direct acting antiviral agents
HCC Hepatocellular carcinoma
HCV Hepatitis C virus
HCW Health care workers
PEG-IFN Pegylated interferon
RBV Ribavirin

REFERENCES

[1] Frank C, Mohamed MK, Strickland GT, Lavanchy D, Arthur RR, Magder LS, El Khoby T, Abdel-Wahab Y, Ohn ESA, Anwar W, Sallam I. The role of parenteral antischistosomal therapy in the spread of hepatitis C virus in Egypt. Lancet 2000;355:887–9.
[2] Kamel MA, Miller FD, El Masry AG, Zakaria S, Khattab M, Essmat G, Ghaffar YA. The epidemiology of Schistosoma mansoni, hepatitis B and hepatitis C infection in Egypt. Ann Trop Med Parasitol 1994;88:501–9.
[3] Kamal S, Madwar M, Bianchi L, Tawil AE, Fawzy R, Peters T, Rasenack JW. Clinical, virological and histopathological features: long-term follow-up in patients with chronic hepatitis C co-infected with S. mansoni. Liver July 2000;20(4):281–9.
[4] Said F, El Beshlawy A, Hamdy M, El Raziky M, Sherif M, Abdel kader A, Ragab L. Intrafamilial transmission of hepatitis C infection in Egyptian multitransfused thalassemia patients. J Trop Pediatr August 2013;59(4):309–13.

[5] Nafeh MA, Medhat A, Shehata M, Mikhail NN, Swifee Y, Abdel-Hamid M, Watts S, Fix AD, Strickland GT, Anwar W, Sallam I. Hepatitis C in a community in upper Egypt: 1. Cross-sectional survey. Am J Trop Med Hyg 2000;63:236–41.

[6] Miller FD, Abu-Raddad LJ. Evidence of intense ongoing endemic transmission of hepatitis C virus in Egypt. Proc Natl Acad Sci USA 2010;107:14757–62.

[7] El-Zanaty F, Way A. Egypt demographic and health survey. Cairo, Egypt: Ministry of Health, El-Zanaty and Associates, and Macro International 2009; 2008.

[8] Kamel MA, Ghaffar YA, Wasef MA, Wright M, Clark LC, Miller FD. High HCV prevalence in Egyptian blood donors. Lancet 1992;340:427.

[9] Habib M, Mohamed MK, Abdel-Aziz F, Magder LS, Abdel-Hamid M, Gamil F, Madkour S, Mikhail NN, Anwar W, Strickland GT, Fix AD, Sallam I. Hepatitis C virus infection in a community in the Nile Delta: risk factors for seropositivity. Hepatology January 2001;33(1):248–53.

[10] Darwish MA, Faris R, Darwish N, Shouman A, Gadallah M, El-Sharkawy MS, Edelman R, Grumbach K, Rao MR, Clemens JD. Hepatitis C and cirrhotic liver disease in the Nile Delta of Egypt: a community-based study. Am J Trop Med Hyg 2001;64:147–53.

[11] El-Zanaty and Associates, Egypt and ICF International. Egypt health Issues survey. Cairo, Egypt and Rockville, MD: Ministry of Health and ICF International 2015; 2015.

[12] Breban R, Doss W, Esmat G, et al. Towards realistic estimates of HCV incidence in Egypt. J Viral Hepat 2013;20:294–6.

[13] Medhat A, Shehata M, Magder LS, Mikhail N, Abdel-Baki L, Nafeh M, Abdel-Hamid M, Strickland GT, Fix AD. Hepatitis c in a community in upper Egypt: risk factors for infection. Am J Trop Med Hyg May 2002;66(5):633–8.

[14] Waked I, Doss W, El-Sayed MH, Estes C, Razavi H, Shiha G, Yosry A, Esmat G. The current and future disease burden of chronic hepatitis C virus infection in Egypt. Arab J Gastroenterol June 2014;15(2):45–52.

[15] Saleh DA, Shebl FM, El-Kamary SS, Magder LS, Allam A, Abdel-Hamid M, Mikhail N, Hashem M, Sharaf S, Stoszek SK, et al. Incidence and risk factors for community-acquired hepatitis C infection from birth to 5 years of age in rural Egyptian children. Trans R Soc Trop Med Hyg 2010;104:357–63.

[16] El-Shanshory MR, Kabbash IA, Soliman HH, Nagy HM, Abdou SH. Prevalence of hepatitis C infection among children with β-thalassaemia major in Mid Delta, Egypt: a single centre study. Trans R Soc Trop Med Hyg 2013;107.

[17] Salama KM, Selim Oel S. Hepatitis G virus infection in multitransfused Egyptian children. Pediatr Hematol Oncol 2009;26(4):232–9.

[18] Abdelwahab MS, El-Raziky MS, Kaddah NA, Abou-Elew HH. Prevalence of hepatitis C virus infection and human immunodeficiency virus in a cohort of Egyptian hemophiliac children. Ann Saudi Med 2012;32(2):200–2.

[19] Khalifa AS, Salem M, Mounir E, El-Tawil MM, El-Sawy M, Abd Al-Aziz MM. abnormal glucose tolerance in Egyptian beta-thalassemic patients: possible association with genotyping. Pediatr Diabetes 2004;5(3):126–32.

[20] Abdalla NM, Galal A, Fatouh AA, Eid K, Salama EEE, Gomma HE. Transfusion transmitted virus (TTV) infection in polytransfused Egyptian thalassemic children. J Med Sci 2006;6(5):833–7.

[21] Omar N, Salama K, Adolf S, El-Saeed GS, Abdel Ghaffar N, Ezzat N. Major risk of blood transfusion in hemolytic anemia patients. Blood Coagul Fibrinolysis 2011;22(4):280–4.

[22] Mansour AK, Aly RM, Abdelrazek SY, Elghannam DM, Abdelaziz SM, Shahine DA, Elmenshawy NM, Darwish AM. Prevalence of HBV and HCV infection among multi-transfused Egyptian thalassemic patients. Hematol Oncol Stem Cell Ther 2012;5(1):54–9.

[23] Kamal AT, Farres MN, Eissa AM, Arafa NA, Abdel-Reheem RS. Incidence of hepatitis c virus seroconversion among hemodialysis patients in the Nile Delta of Egypt: a single-center study. Saudi J Kidney Dis Transpl 2017 January–February;28(1):107–14.

[24] Gohar SA, Khalil RY, Elaish NM, Khedr EM, Ahmed MS. Prevalence of antibodies to hepatitis C virus in hemodialysis patients and renal transplant recipients. J Egypt Public Health Assoc 1995;70(5–6):465–84.

[25] Senosy SA, El Shabrawy EM. Hepatitis C virus in patients on regular hemodialysis in Beni-Suef Governorate. Egypt J Egypt Public Health Assoc June 2016;91(2):86–9.

[26] Saleh DA, Shebl F, Abdel-Hamid M, Narooz S, Mikhail N, El-Batanony M, El-Kafrawy S, El-Daly M, Sharaf S, Hashem M, et al. Incidence and risk factors for hepatitis C infection in a cohort of women in rural Egypt. Trans R Soc Trop Med Hyg 2008;102:921–8.

[27] Lohiniva AL, Talaat M, Bodenschatz C, Kandeel A, El-Adawy M, Earhart K, Mahoney FJ. Therapeutic injections in the context of Egyptian culture. Promot Educ 2005;12(1):13–8.

[28] Talaat M, Kandeel A, El-Shoubary W, Bodenschatz C, Khairy I, Oun S, Mahoney FJ. Occupational exposure to needlestick injuries and hepatitis B vaccination coverage among health care workers in Egypt. Am J Infect Control December 2003;31(8):469–74.

[29] Shalaby S, Kabbash IA, El Saleet G, Mansour N, Omar A, El Nawawy A. Hepatitis B and C viral infection: prevalence, knowledge, attitude and practice among barbers and clients in Gharbia governorate. Egypt East Mediterr Health J 2010;16(1):10–7.

[30] El-Ghazzawi E, Drew L, Hamdy L, El-Sherbini E, Sadek Sel D, Saleh E. Intravenous drug addicts: a high risk group for infection with human immunodeficiency virus, hepatitis viruses, cytomegalovirus and bacterial infections in Alexandria Egypt. J Egypt Public Health Assoc 1995;70(1–2):127–50.

[31] Sievert W, Altraif I, Razavi HA, Abdo A, Ahmed EA, Alomair A, Amarapurkar D, Chen CH, Dou X, El Khayat H, et al. A systematic review of hepatitis C virus epidemiology in Asia, Australia and Egypt. Liver Int 2011;31(Suppl. 2):61–80.

[32] Aceijas C, Rhodes T. Global estimates of prevalence of HCV infection among injecting drug users. Int J Drug Policy 2007;18(5):352–8.

[33] Lehman EM, Wilson ML. Epidemic hepatitis C virus infection in Egypt: estimates of past incidence and future morbidity and mortality. J Viral Hepat 2009;16:650–8.

[34] Miller FD, Abu-Raddad LJ. Quantifying current hepatitis C virus incidence in Egypt. J Viral Hepat September 2013;20(9):666–7.

[35] Kamal SM. Acute hepatitis C: a systematic review. Am J Gastroenterol May 2008;103(5):1283–97.

[36] Kamal SM. Acute hepatitis C: prospects and challenges. World J Gastroenterol December 28, 2007;13(48):6455–7.

[37] Esmat G, Hashem M, El-Raziky M, et al. Risk factors for hepatitis C virus acquisition and predictors of persistence among Egyptian children. Liver Int 2012;32:449–56.

[38] Thomas DL, Seeff LB. Natural history of hepatitis C. Clin Liver Dis 2005;9:383–98.

[39] Heller T, Rehermann B. Acute hepatitis C: a multifaceted disease. Semin Liver Dis 2005;25:7–17.

[40] Chung RT. Acute hepatitis C virus infection. Clin Infect Dis 2005;41:S14–7.

[41] Irving WL. Acute hepatitis C virus infection: a neglected disease?. Gut 2006;55:1075–7. Dis 2005;41:S14–S17.

[42] Meky FA, Stoszek SK, Abdel-Hamid M, et al. Active surveillance for acute viral hepatitis in rural villages in the Nile Delta. Clin Infect Dis 2006;42:628–33.

[43] Zakaria S, Fouad R, Shaker O, et al. Changing patterns of acute viral hepatitis at a major urban referral center in Egypt. Clin Infect Dis 2007;44:e30–6.

[44] Mohamed MK, Abdel-Hamid M, Mikhail NN, et al. Interfamilial transmission of hepatitis C in Egypt. Hepatology 2005;42:683–7.

[45] Magder LS, Fix AD, Mikhail NN, et al. Estimation of the risk of transmission of hepatitis C between spouses in Egypt based on seroprevalence data. Int J Epidemiol 2005;34:160–5.

[46] Kamal SM, Amin A, Madwar M, et al. Cellular immune responses in seronegative sexual contacts of acute hepatitis C patients. J Virol 2004;78:12252–8.

[47] Sulkowski MS, Ray SC, Thomas DL. Needlestick transmission of hepatitis C. JAMA 2002;287:2406–13.

[48] Kubitschke A, Bader C, Tillmann HL, et al. Injuries from needles contaminated with hepatitis C virus: How high is the risk of seroconversion for medical personnel really? Internist 2007;48:1165–72.

[49] Mehta A, Rodrigues C, Ghag S, et al. Needlestick injuries in a tertiary care centre in Mumbai, India. J Hosp Infec 2005;60:368–73.

[50] Santilli C. Medical care, screening and regularization of Sub-Saharan Irregular migrants affected by hepatitis B in France and Italy. J Immigr Minor Health April 20, 2017.

[51] Daw MA, El-Bouzedi A, Ahmed MO, Dau AA, Agnan MM. Hepatitis C virus in North Africa; an emerging state. ScientificWorldJournal 2016;2016:7370524.

[52] Daw MA, El-Bouzedi A, Ahmed MO, Dau AA, Agnan MM. Epidemiology of hepatitis C virus and genotype distribution in immigrants crossing to Europe from north and sub- Saharan Africa. Trav Med Infect Dis 2016:15.

[53] Estes C, Abdel-Kareem M, Abdel-Razek W, Abdel-Sameea E, Abuzeid M, Gomaa A, Osman W, Razavi H, Zaghla H, Waked I. Economic burden of hepatitis C in Egypt: the future impact of highly effective therapies. Aliment Pharmacol Ther September 2015;42(6):696–706.

Chapter 3.2

Hepatitis C in North Africa (Arabic Maghreb Region)

Mohamed A. Daw
University of Tripoli, Tripoli, Libya

Chapter Outline

INTRODUCTION

Hepatitis C virus (HCV) infection has received great attention worldwide as it is associated with a high morbidity and mortality. No country rich or poor is safe from this infection. It has been greatly reflected on all aspects of life, socially, economically, and even politically. It is a dynamic infection. The prevalence of the virus varies greatly from one country to another and even within the regions of the same country. The burdens of HCV infection lie heavily particularly on low-income countries [1]. HCV infection is a major public health concern among African countries in particular; they have the highest prevalence rates of HCV in the world, ranging from 1% to 26% [2,3]. Over 28 million people are chronically infected with HCV in this continent,

Hepatitis C in Developing Countries. https://doi.org/10.1016/B978-0-12-803233-6.00006-0

and it is difficult to speculate about current and future trends [4]. In Africa the prevalence of HCV varies from one country to another. Comparison of the epidemiologic status in different countries poses some difficulties because of variations in diagnostic procedures, various definitions of infection, different methodologies used, and the time when the epidemiologic research took place [5]. The Global Burden of Diseases project subdivides the African continent into four macro areas: Central, East, West, and Southern, whereas the Saharan area (North Africa) is generally associated with Middle East countries. The estimated prevalence of anti-HCV in the whole Sub-Saharan Africa is 2.9% with an estimated 26.9 million cases, ranging from 6.0% in the Central area to 0.9% in the Southern countries. The viremic rate is estimated at 70.5%, with a peak of 79.6% in Western Africa, accounting to almost 20.0 millions active HCV replication cases [6].

Although, there is much in common among these countries with regard to the environment, population genetics, habitual, and language, they differ greatly in the infrastructure of health care system, personnel income, level of urbanization, and degree of poverty. The population varies from poor nomadic Bedouins scattered in Sahara area of Mauritania, Algeria, and Morocco to sheltered valleys in the Atlas Mountains, up to the Mediterranean coast cities of Libya. Epidemiologic information regarding HCV among these population is sparse and fragmentary and further hampered by poverty, ignorance, and more recently uprising conflicts [7].

THE STATUS OF HCV IN MAGHREB REGION

Accurate data on the burden of hepatitis C infection in this region are hampered by the lack of adequate surveillance and poor resources for proper data collection and management. Despite the geographic proximity and social and heritage interaction between these countries, the prevalence and complication of HCV are greatly variable. A part from Libya and Morocco, most states of the region lack adequate surveillance studies on HCV to enable them to take evidence-based policy decisions [8,9]. Few studies have been carried out within these countries. The data are outdated, aggregated, or limited to a small specific population affected by a selection bias because most studies are generally based on subjects from risk groups or blood donors with information lacking on children and senior citizens who are generally not included as specialized groups [10,11]. In Maghreb region, predisposing factors and epidemiologic information of HCV vary significantly from one country to another and considerable variation was evident in economic and geographical entity of these countries as shown in Table 3.2.1. The region's contribution to global literature regarding HCV remains relatively small. Though there has been an increase in the publications from Libya and Morocco, others remain to follow.

TABLE 3.2.1 Development Parameters and Genotype Distribution of HCV Among Each Country of Maghreb Region

Countries	Libya	Tunis	Algeria	Morocco	Mauritania
		Maghreb Region Countries			
Country's population (in millions)	6.4200	11	35	32	3.5
Population density/km	04	71	16	76	04
GNI (US$ as PPP)	12.020	3720	4420	2770	2400
HDI (2010–14)	0.849	0.712	0.736	0.628	0.453
Country classification	HI	LMI	LMI	LMI	LI
Genotypes (%)					
1 1a 1b	35% 1a 1b	67% 1a 1b	89% 1a 1b	68% 1a 1b	?
2	14.2	13%	9%	30%	?
2a	2a	2a	2a,b	2i,2k	
3	15%	3%	1%	–	?
4	29.2%	21%	1%	–	?
4a	4a	4a	4a	–	
4k	4k	4k	4c,d	–	?
5	0.2%	–	–	–	?

The estimated prevalence of HCV in the whole region is 2.2%, ranging between 0.6% and 4.8%, depending on the country. According to current accurate estimates, the lowest prevalence of the virus is in Libya and the highest is in Mauritania and Algeria. It is considered to be the lowest among other African countries and Middle east regions including the Arabian peninsula and Sham region. The HCV viremic rate is 67.8%, with 1.9 million active HCV replication cases. In this region, genotype distribution is highly heterogenic. The predominant genotype is the genotype 1, followed by 4 and 3. Only small percentages of genotype 2 and 5 and no genotype 6 were reported.

PREVALENCE OF HCV IN MAGHREB COUNTRIES

HCV in Algeria

Algeria is the largest country of Africa, with populations living a range of traditional/rural and modern/urban lifestyles. There is scarcity of data regarding the epidemiology of HCV in Algeria. Official data suggest that distinct hepatitis C epidemics are currently affecting Algerian population [12]. The Algerian Ministry of Health estimated that the prevalence of HCV infection had reached 2.5% in Algeria [13,14]. In a rare retrospective study by Rouabhia et al., HCV markers were detected in 739 diabetic and 580 nondiabetic patients attending the Internal Medicine Department of University Hospital Center of Batna-Algeria [15]. Anti-HCV seropositivity was 17.5% in diabetic patients and 8.4% in nondiabetic patients ($P<.01$). However, after adjustment for age, this difference is statistically significant only in patients aged between 40 and 65 years (22.2% vs. 9.3%, $P=.024$). Despite the ongoing argument whether diabetes mellitus is a risk factor for HCV infection; or is HCV a risk factor for type 2 diabetes mellitus, the prevalence of HCV among the healthy Algerian in this group is indeed high as it ranged from 8.4% to 9.3%. Further studies are needed to clarify such conclusion.

High HCV prevalence rates reaching 53% were reported in patients undergoing hemodialysis and 31.6% among hemophiliacs in Algeria. This situation is plausibly connected with nosocomial transmission and occupational exposure to HCV among health care workers [14]. In Algeria HCV genotype 1 was the most frequent (88.7%), followed by genotypes 2 (8.5%), 4 (1.1%), 3 (0.9%), and 5 (0.2%). The genotype distribution was related to age and region. Genotype 1 was significantly less frequent in the ≥60 age group than in the younger age group [odds ratio (OR)=0.2; 95% confidence interval (CI): 0.1–0.5, $P<.001$] [16,17]. Furthermore, genotype 1 was more frequent in the central part of North Eastern Region of Algeria than elsewhere (Fig. 3.2.1).

HCV in Libya

Libya is the second largest and richest country in North Africa with a small population. HCV is well documented in this country, and major studies were carried out concerning all aspects of HC infections over the last 20 years [18]. The largest comprehensive national study in the developing countries of Asia and Africa was carried in Libya covering over 1% of the total population [8]. Accordingly the overall study, the prevalence rates of HCV in males and females were similar (1.1% and 1.3%, respectively). The mean age of anti-HCV–positive females was 31.7 ± 18.4 years and that of males was $35.6+20.9$ years). The mean age of HCV-positive individuals was significantly higher than that of anti-HCV–negative individuals for both males (almost 10 years difference) and females (7 years difference). HCV was more prevalent among single and younger individuals and about 40% of HCV cases were also below 30 years old. The prevalence

Arab Maghreb Union (AMU)

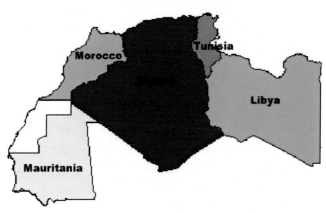

FIGURE 3.2.1   The Arab Maghreb Region.

of HCV was higher among the illiterate group (3.1%), whereas in the literate groups, it ranged from 0.9% to 1.1%. Hepatitis C was most prevalent among intravenous drug users (IVDUs) (7.4%) and less prevalent but still substantial in those undergoing blood transfusion (2.7%), surgical operation (2.3%), or hospital admission (1.9%).

The prevalence of HCV infection in Libya varied widely between HD centers from 0% to 75.9%. Seropositive patients were younger and had been receiving dialysis for substantially longer. Follow-up revealed an incidence of sero-conversion of 7.1% during the first year of dialysis [19]. A total of 20 discrete genotypes and subtypes were identified among the Libyan population ranging from 11.5% to 0.3% across the country. Genotype 1 was the most frequent among all regions (19.7% to 40.5%), reaching the highest value in Tripoli region, followed by genotype 4, which was more prevalent in the South (49.3%) and West (40.0%) regions. Genotype 3 was higher in Tripoli (21.3%) and East (15.9%) regions, whereas genotype 2 was common in North (23.6%) and South (22.5%) regions [15,19]. The frequency of these genotypes is significantly associated with demographic and risk factors involved. Transmission by IVDUs is the most common route of transmission in Libya and has become more frequent and associated with genotype 1 (49.2%) and genotype 3 (32.6%) infections (Fig. 3.2.2). The prevalence of HBV, HCV and HIV is shown in Fig. 3.2.3.

HCV in Mauritania

Mauritania is a Saharan country characterized by a very high prevalence of viral hepatitis, which poses a very serious public health problem. In recent coherent studies, up to 20% of consulting patients, pregnant women, or blood donors have HBV and up to 33% have HDV (hepatitis D virus) [20]. Despite

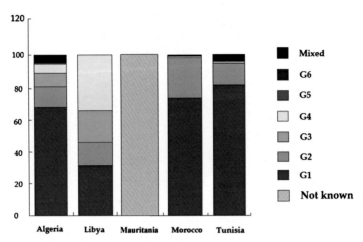

FIGURE 3.2.2 HCV genotypes in Arab Maghreb.

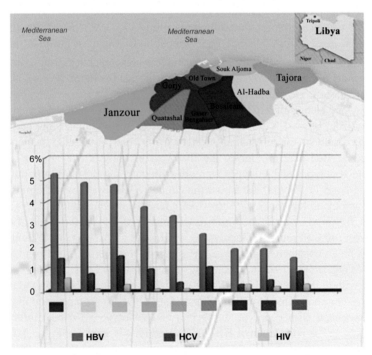

FIGURE 3.2.3 The prevalence of HBV, HCV, and HIV in Libya.

the limited published data regarding HCV in this country, the epidemiologic analysis of blood-borne hepatitis suggests a higher prevalence of HCV among Mauritanians even comparable with other Maghreb countries. A comprehensive prospective study carried by Mansour et al., on a total of 1966 subjects, showed that the overall prevalence of HBsAg was 18.3%. It was significantly higher in males (24.4%) than in females (13.8%) [$P<.001$ (OR = 2.04 (1.46–2.85), $P<.001$] and varied significantly over the different ethnic groups: 22.7% in White Moor, 19.7% in Black Moor, and 12% in African ethnicities [$P = 0.025$, OR = 0.47 (0.26–0.82), $P = 0.008$ for the comparison between White Moor and other African ethnic groups] [21,22]. The characteristics of the positive subjects strongly suggest healthcare-associated transmission, intrafamilial transmission, sexual transmission, and more frequently, a history of hospitalization, and transmission through iatrogenic, medical, or paramedical procedures.

In 2000, a total of 2854 healthy blood donors were screened for HCV antibodies in the National Hospital of Nouakchott, the prevalence of HCV was found to be 2.7%; however, no risk factors were studied [23,24]. However, speculation has been raised that the prevalence of HCV in Mauritania may reach up to 10.7 mimics that of west African countries hence then, further studies are needed [25].

HCV in Morocco

Morocco is the bounding arm of the Maghreb region facing both the Atlantic Ocean from West and the Mediterranean Sea from the North. Different studies were carried out regarding HCV in Morocco both in general population and on higher risk groups. Early studies indicated that the prevalence of HCV was 1.93% among general population, though it was 1.08% within blood donors [26]. Recently, a nationwide cross-sectional survey carried out in 11 major Moroccan regions within a 6-year period showed that the overall prevalence of HCV infection in the general population was found to be 1.58%, though it was less among blood donors [9]. The prevalence was higher among males aged less than 30 years. Factors significantly associated with HCV infection were increasing age, dental treatment, use of glass syringes, and surgical history. Emerging data suggest that differences may exist in the prevalence of anti-HCV between northern and southern regions in Morocco [27]. HCV is a major problem among hemodialysis centers in Morocco. A multicenter study in different Moroccan dialysis centers found that prevalence varied between the centers from 11% to 91% [28].

Traditional medicine and barbers play an important role the spread of HCV within Moroccan community. A survey carried out on barbers and their clients indicated that they hover over 5% of anti-HCV as they were working in unsanitary conditions [16]. However, this poses a particular problem for the Maghreb region, particularly Libya as most of the hairdressers are Moroccans. Drug addiction is a serious problem in Morocco and has been considered as a "male-associated habit" as the overwhelming majority were young males who were either single or divorced. HCV is highly "seroprevalent" among this group, reaching up to 60% [29]. The commonly reported HCV genotype in Morocco were genotype 1 (46%) and 2 (40%), followed by genotypes 3 and 4, though in injecting drug users (IDUs) genotype 1 accounted for 65% followed by genotype 3 (26%) and 4 (10%) [30].

HCV in Tunisia

Tunisia is the smallest geographical area among the Maghreb countries bounded by the two largest countries in the region—Libya and Algeria. Different studies were carried on HCV infection in Tunis. Two decades ago, Tirki et al. had performed a study on the prevalence of viral hepatitis in Tunis including hepatitis B, C, and delta virus on a selective population composed of mainly male military recruits aged 20–25 years [17]. The overall prevalence of HDV and HCV was 17.7, though it was low for HCV alone. In 2005 another study had found that the prevalence of HCV reached up to 1.7% among the general population with great heterogeneity in the geographical distribution of HCV in Tunisia. HCV was particularly higher in the northwestern region of the country rather than anywhere else, though there was no difference in positivity according to gender or living in rural or urban settings; the only significant risk factor was advanced age [31]. Similar results were reported among blood donors and diabetic patients as anti-HCV antibodies varied between 0.5% and 1.8% [32]. However, these studies were lavished by a lack of specificity and confined to a certain population. They did not a really mirror the accurate status of HCV within the country. Recently, a seroprevalency study of Transfusion Transmitted Infections in First-Time Volunteer and Replacement Donors in Tunisia showed that HCV in adjusted model reached 1.9 (95% CI: 0.9–4.1 $P = .11$). Hence, further studies are needed to assess the actual burden of CV infection in Tunis [33].

The prevalence of HCV infection among Tunisian dialysis patients was reported to be high, reaching up to 51%. There was a close correlation between the number of anti-HCV–positive patients and the duration on dialysis [34,35]. This highlights the nosocomial transmission of HCV in dialysis units where the number of infected patients is high and where the material management does not take into account the patient's viral status. The genotyping patterns among Tunisian's varies greatly from one study to another. Subtype 1b was largely predominant (79%), whereas types 1a, 2a, 2b, 3a, and 4a occurred much less frequently.

FACTORS ASSOCIATED WITH HCV IN MAGHREB REGION

Health Care–Associated Factors

Hemodialysis, blood transfusion, and hospital care–associated practice combined with habitual and community-associated factors were among the most important risk factors for HCV in Maghreb countries. The mortality and morbidity of the infection associated with such factors are expected to be high and the adjusted hazard ratio for patient death may reach 2.3 [36]. The prevalence of HCV infection observed in blood products and dialysis patients in these countries are much higher than those corresponding in the general population. This is further deteriorated in areas of chronic conflict, notably Libya, whose blood supplies were safer and more secure before the 2011 uprisings. Tunisia and Libya experienced stock-outs during their recent uprisings both for blood-screening reagents and access to new antiretroviral drugs, hence evaluation of the impact of the conflicts on the health care system are needed [37]. Health care workers should be trained to familiarize themselves with the risk of acquiring HCV via sharps injury and other nosocomial routes. Habitual and cultural factors that may influence the spread of HCV in Maghreb countries include male and female circumcision, particularly in Mauritania and Morocco. Hijiama done by an informal practitioner, tattooing, folk body piercing and threading, sharing hygiene and sharp items, and the use of communal barber may be considered as risk factors for HCV particularly among rural dwellers [23,34]. Education and public communications in highlighting such malpractice and habits s are needed.

Illegal Migration, Drug Trafficking and Abusing

The Arab Maghreb region is a major starting point for illelgal migration from the South towards Europe (Fig. 3.2.4). Migration contributes to transmission of HCV (hepatitis B virus), HBV or HIV from endemic areas in sub-Saharan Africa to the Arab Maghreb particularly Libya. Drug trafficking, IVDUs, and sniffing pose specific problems among the countries of Maghreb region. This is enhanced and further complicated by the geographical location of the vast area of the region. Morocco is the world's foremost producer of cannabis resin and remains the main source of the drug for the consumer market in Western Europe and the largest seizures occurred in 2007 was in Mauritania [38,39]. Studies from MENA regions indicated that after Iran, the largest numbers of people injecting drugs are from Egypt and Algeria [40]. Injecting drug use is also a significant route of HIV transmission in Libya and Tunisia. Cocaine use is reportedly increasing in these countries, but as yet cocaine injecting is not reported. Noninjecting drug use has not been highlighted as linked to the

Top three nationalities of migrants using the most popular routes

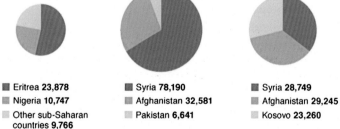

C Mediterranean	E Mediterranean	W Balkan
■ Eritrea **23,878**	■ Syria **78,190**	■ Syria **28,749**
■ Nigeria **10,747**	■ Afghanistan **32,581**	■ Afghanistan **29,245**
■ Other sub-Saharan countries **9,766**	■ Pakistan **6,641**	■ Kosovo **23,260**

FIGURE 3.2.4   Illegal migration patterns.

transmission of either HIV or HCV in North African countries. Information on HCV prevalence among people who inject drugs is not available for most countries in the region, but, where reported, the data reveal high levels of HCV [41]. Furthermore, men who have sex with men and people who inject drugs, both highly criminalized populations in region, are more affected by HCV and HIV.

Level of Urbanization

Although there is no universally accepted definition of what urban is? Urbanization has contributed to an overall improvement of health [42]. Different studies have shown that the level of urbanization could influence the prevalence of HCV in Maghreb region. Libya, which is considered to be an urbanized country, has the lowest HCV prevalence. HCV is also prevalent in areas which are mainly rural and those regions that are not expected to pass the tipping point before 2030 [43]. Sociodemographic (rural vs. urban) studies on population within endemic countries have showed a higher prevalence in rural population than urban population [44]. Blood donors and children from rural areas had a higher prevalence of HCV than those from rural areas. Results obtained from Tunis, Algeria, and Morocco showed that HCV was significantly higher in the north-western region and suburban area of Tunis than in the northern region. In Algeria, HCV was highly associated with population living in remote mountains or desert regions than the urbanized populations living in Mediterranean coast [16]. In Morocco, being a rural resident was found to be greatly associated with HCV infection, whereas in Mauritania, the vast majority of residents are either rural dwelling or Bedouins even with the capital cities. This could be contributed to the habitual and economic status of this population. Lack of sanitary services, literacy, traditional medicine, occupational hazards, and lack of adapting safe medical practice may act as influencing factors. However, further studies are needed to clarify such an assumption.

STRATEGIES TO PREVENT HCV IN THE MAGHREB REGION

The status of HCV among Arabic Maghreb countries should be a worrying issue to all sectors involved in public health, particularly those who are in close contact with patients and strategists who should plan for new healthy look of the Arabic-African societies [45]. Major measures should be implemented. These include immediate primary intervention (short-run) strategies and long-run prevention strategies. Such prevention strategies should target reduction of transmission of HCV, particularly among those at risk of acquiring the virus. Risk reduction counseling and HCV-screening programs were directed to specific population as suggested by the Centers for Disease Control, which may include persons on long-term dialysis, chronic liver disease patients, health-care workers after needlestick injury, and children of HCV-positive mothers. Preventive measures should be directed toward individuals or populations at specific settings such IDUs/STD and prisoners. Hospitals and health care centers should adapt universal and specific infection control programs targeting not only the status of nosocomial infection per se but also particularly those units or persons who were more prone to HCV infection.

Specific prevention programs should take place by adapting better cleansing and standard sterilization methods to stop nosocomial and iatrogenic transmission of HCV. Patient care practices associated with higher HCV prevalence among chronic hemodialysis units should be identified, and recommendations and precautions in these settings should be adapted. This includes routinely wearing gloves; restricting the use of common supplies, instruments, and medications for multiple patients; prohibiting the use of mobile carts within treatment areas to store or distribute medications and clean supplies. Further programs should include obligatory advanced laboratory-screening methods for blood and blood products and reduction of the number of transfusion-related transmissions and promote judicious injection and parenteral medications among doctors and patients. Once patients are found to have hepatitis C, they need to be counseled and clinically evaluated to reduce the risk of transmission and stop the progression of the disease, respectively. Patients with chronic HCV who are susceptible to HAV or HBV infections should be vaccinated, particularly those among the Arab countries as the prevalence of these two viruses is high in this region. Furthermore, HCV is considered to be an opportunistic disease in persons with HIV infection. Hence, the safety and efficacy of antiviral therapy using INF-& and/or INF-& plus ribavirin for HCV coinfected patients must be evaluated scientifically through clinic trial in the region. The implementation of new antiviral treatment and the role of genotyping also have to be assessed, and appropriate strategies for the management of end-stage liver diseases have to be rigorously investigated.

CONCLUSION

HCV is a serious on-going problem in Maghreb countries, and it has great socialeconomic impact, which may touch the future of the young generation and influence the infrastructure of such dynamic states. Despite the great human and natural resources that this nation has, none of these countries have shown meticulous and clear national or regional scientific plans to combat the future damage that HCV may cause. Prevalence of HCV is destined to increase further among Arab Maghreb region.

ABBREVIATION

IVDUs Intravenous drug users

REFERENCES

[1] Daw MA, Dau AA. Hepatitis C virus in Arab world: a state of concern. Sci J 2012;2012:719494. http://dx.doi.org/10.1100/2012/719494. PMID:22629189.

[2] Lazarus J. Global policy report on the prevention and control of viral hepatitis in WHO Member States. World Health Organization; 2014.

[3] Madhava V, Burgess C, Drucker E. Epidemiology of chronic hepatitis C virus infection in sub Saharan Africa. Lancet Infect Dis May 31, 2002;2(5):293–302.

[4] Onyekwere CA, Hameed L. Hepatitis B and C virus prevalence and association with demographics: report of population screening in Nigeria. Trop Doctor October 1, 2015; 45(4):231–5.

[5] Te HS, Jensen DM. Epidemiology of hepatitis B and C viruses: a global overview. Clin Liver Disease February 28, 2010;14(1):1–21.

[6] Petruzziello A, Marigliano S, Loquercio G, Cozzolino A, Cacciapuoti C. Global epidemiology of hepatitis C virus infection: an update of the distribution and circulation of hepatitis C virus genotypes. World J Gastroenterol 2016;22(34):7824–40.

[7] Daw MA, El Bouzedi A, Dau AA. Libyan armed conflict 2011: mortality, injury and population displacement. Afr J Emerg Med September 30, 2015;5(3):101–7.

[8] Daw MA, El Bouzedi A. In association with Libyan Study Group of Hepatitis & HIV. Prevalence of hepatitis B and hepatitis C infection in Libya: results from a national population based survey. BMC Infect Dis 2014;14:17.

[9] Baha W, Foullous A, Dersi N, They they TP, Nourichafi N, Oukkache B, Lazar F, Benjelloun S, Ennaji MM, Elmalki A, Mifdal H. Prevalence and risk factors of hepatitis B and C virus infections among the general population and blood donors in Morocco. BMC Public Health January 18, 2013;13(1):1.

[10] Daw MA. Transmission of hepatitis C virus. In: Stambouli O, editor. Hepatitis C virus: molecular pathways and treatments. Nevada, USA: OMICS Group; 2014.

[11] Daw MA, Elkaber MA, Drah AM, Werfalli MM, Mihat AA, Siala IM. Prevalence of hepatitis C virus antibodies among different populations of relative and attributable risk. Saudi Med J 2002;23(11):1356–60.

[12] Bensalem A, Selmani K, Narjes H, Bencherifa N, Mostefaoui F, Kerioui C, Pineau P, Debzi N, Berkane S. Eastern region represents a worrying cluster of active hepatitis C in Algeria in 2012. J Med Virology February 1, 2016.

[13] National Travel Health Network Centre. Country information, Algeria, Tunisia, Morocco. 2009.

[14] Ayed Z, Houinato D, Hocine M, Ranger Rogez S, Denis F. Prevalence of serum markers of hepatitis B and C in blood donors and pregnant women in Algeria. B Soc Pathol Exot 1995;88:225–8.

[15] Rouabhia S, Sadelaoud M, Chaabna Mokrane K, Toumi W, Abenavoli L. Hepatitis C virus genotypes in north eastern Algeria: a retrospective study. World J Hepatol 2013;5(7):393–7. http://dx.doi.org/10.4254/wjh.v5.i7.393. Available from: http://www.wjgnet.com/1948.5182/full/mv5/i7/393.htm.

[16] Belbacha I, Cherkaoui I, Akrim M, Dooley KE, El Aouad R. Seroprevalence of hepatitis B and C among barbers and their clients in the Rabat region of Morocco. East Mediterr Health J December 1, 2011;17(12):911.

[17] Triki H, Said N, Ben Salah A, Arrouji A, Ben Ahmed F, Bouguerra A, Hmida S, Dhahri R, Dellagi K. Seroepidemiology of hepatitis B, C and delta viruses in Tunisia. Trans R Soc Trop Med Hyg 1997;91:11–4.

[18] Elasifer HA, Agnnyia YM, Al Alagi BA, Daw MA. epidemiological manifestations of hepatitis C virus genotypes and its association with potential risk factors among Libyan patients. Virol J 2010;7:317.

[19] Daw MA, El Bouzedi A, Dau AA. Geographic distribution of HCV genotypes in Libya and analysis of risk factors involved in their transmission. BMC Res Notes 2015;8:367. http://dx.doi.org/10.1186/s13104.015.1310.x. PMID:26293137.

[20] Lunel Fabiani F, Mansour W, Amar AO, Aye M, Le Gal F, Malick FZ, Baïdy L, Brichler S, Veillon P, Ducancelle A, Gordien E. Impact of hepatitis B and delta virus co infection on liver disease in Mauritania: a cross sectional study. J Infect November 30, 2013;67(5):448–57.

[21] Mansour W, Malick FZ, Sidiya A, Ishagh E, Chekaraou MA, Veillon P, et al. Prevalence, risk factors, and molecular epidemiology of hepatitis B and hepatitis delta virus in pregnant women and in patients in Mauritania. J Med Virol 2012;84(8):1186e98.

[22] Mansour W, Bollahi MA, Hamed CT, Brichler S, Le Gal F, Ducancelle A, et al. Virological and epidemiological features of hepatitis delta infection among blood donors in Nouakchott, Mauritania. J Clin Virol 2012;55(1):12e6.

[23] Aye MO, Moktar MM. Épidémiologie de l'hépatite C en Mauritanie. J Afr D'Hépato Gastroentérologie December 1, 2007;1(3):141–2.

[24] Lo BB, Meymouna M, Boulahi MA, Tew M, Sow A, Ba A, Sow MB. Prévalence des marqueurs sériques des virus des hépatites B et C chez les donneurs de sang à Nouakchott, Mauritanie. Bull Soc Pathol Exot 1999;92:83–4.

[25] Riou J, Aït Ahmed M, Blake A, Vozlinsky S, Brichler S, Eholié S, Boëlle PY, Fontanet A. Hepatitis C virus seroprevalence in adults in Africa: a systematic review and meta-analysis. J Viral Hepatitis October 2015.

[26] Benjelloun S, Bahbouhi B, Sekkat S, Bennani A, Hda N, Benslimane A. Anti HCV seroprevalence and risk factors of hepatitis C virus infection in Moroccan population groups. Res Virol 1996;147:247–55.

[27] Boutayeb H, Aamoum A, Benchemsi N. Connaissances sur les virus des hepatites B et C et le VIH chez des donneurs de sang a Casablanca. East Mediterr Health J September 1, 2006;12(5):538.

[28] Abdelaali B, Omar M, Taoufik D, Samir A, Saad M. Hepatitis C viral prevalence and seroconversion in Moroccan hemodialysis units: eight year follow up. J Med Diagnostic Methods November 20, 2013;2013.

[29] Trimbitas RD, Serghini FZ, Lazaar F, Baha W, Foullous A, Essalhi M, El Malki A, Bellefquih AM, Bennani A. The "hidden" epidemic: a snapshot of Moroccan intravenous drug users. Virol J March 6, 2014;11(1):1.

[30] Brahim I, Akil A, Mtairag EM, Pouillot R, El Malki A, Nadir S, Alaoui R, Njouom R, Pineau P, Ezzikouri S, Benjelloun S. Morocco underwent a drift of circulating hepatitis C virus subtypes in recent decades. Arch Virol March 1, 2012;157(3):515–20.

[31] Mejri S, Salah AB, Triki H, Alaya NB, Djebbi A, Dellagi K. Contrasting patterns of hepatitis C virus infection in two regions from Tunisia. J Med Virol June 1, 2005;76(2):185.

[32] Kaabia N, Ben Jazia E, Slim I, Fodha I, Hachfi W, Gaha R, Khalifa M, Hadj Kilani A, Trabelsi H, Abdelaziz A, Bahri F, Letaief A. Association of hepatitis C virus infection and diabetes in central Tunisia. World J Gastroenterol 2009;15(22):2778–81. http://dx.doi.org/10.3748/wjg.15.2778. Available from: http://www.wjgnet.com/1007 9327/15/2778.asp.

[33] Jemia RB, Gouider E. Seroprevalency of transfusion transmitted infections in first time volunteer and replacement donors in Tunisia. Transfus Clinique Biol December 31, 2014;21(6):303–8.

[34] Sassi F, Gorgi Y, Ayed K, Abdallah TB, Lamouchi A, Maiz HB. Hepatitis C virus antibodies in dialysis patients in Tunisia: a single center study. Saudi J Kidney Dis Transplant April 1, 2000;11(2):218.

[35] Debbeche R, Said Y, Temime BH, El Jery K, Bouzaidi S, Salem M, Najjar T. Epidémiologie de l'hépatite C en Tunisie. La Tunisie Médicale 2013;91:86–91.

[36] Daw MA, Dau AA, Agnan MM. Influence of healthcare associated factors on the efficacy of hepatitis C therapy. Sci World J 2012;2012:580216.

[37] Daw MA, El Bouzedi A, Dau AA. The assessment of efficiency and coordination within the Libyan health care system during the armed conflict 2011. Clin Epidemiol Glob Health August 24, 2015. http://dx.doi.org/10.1016/j.cegh.2015.07.004.

[38] United Nations Office on Drugs, Crime. World drug report. 2007.

[39] Cook C, Kanaef N. The global state of harm reduction 2008: mapping the response to drug related HIV and hepatitis C epidemics. International Harm Reduction Association; 2008.

[40] Nelson PK, Mathers BM, Cowie B, Hagan H, Des Jarlais D, Horyniak D, Degenhardt L. Global epidemiology of hepatitis B and hepatitis C in people who inject drugs: results of systematic reviews. Lancet August 19, 2011;378(9791):571–83.

[41] Mumtaz GR, Weiss HA, Thomas SL, Riome S, Setayesh H, Riedner G, Semini I, Tawil O, Akala FA, Wilson D, Abu Raddad LJ. HIV among people who inject drugs in the Middle East and North Africa: systematic review and data synthesis. PLoS Med June 17, 2014;11(6):e100166344.

[42] Alirol E, Getaz L, Stoll B, Chappuis F, Loutan L. Urbanisation and infectious diseases in a globalised world. Lancet Infect Dis February 28, 2011;11(2):131–41.

[43] Malik A, Awadallah B. The economics of the Arab spring. World Dev May 31, 2013;45:296–313.

[44] Daw MA, El Bouzedi A, Ahmed MO, Dau AA, Agnan MM. Hepatitis C virus in North Africa: an emerging threat. Sci World J August 16, 2016;2016.

[45] Daw MA, El Bouzedi A, Ahmed MO, Dau AA, Agnan MM, Drah AM, Deake AO. Prevalence of human immune deficiency virus in immigrants crossing to Europe from North and Sub Saharan Africa. Trav Med Infect Dis October 21, 2016.

FURTHER READING

[1] Djebbi A, Triki H, Bahri O, Cheikh I, Sadraoui A, Ben Ammar A, Dellagi K. Genotypes of hepatitis C virus circulating in Tunisia. Epidemiol Infect 2003;130:501–5.

[2] Kchouk FH, Gorgi Y, Bouslama L, Sfar I, Ayari R, Khiri H, Halfon P, Aouadi H, Jendoubi Ayed S, Ayed K, Ben Abdallah T. Phylogenetic analysis of isolated HCV strains from Tunisian hemodialysis patients. Viral Immunol February 1, 2013;26(1):40–8.

[3] Daw MA, El Bouzedi A, Ahmed MO, Dau AA, Agnan MM. Hepatitis C virus in North Africa; an emerging state. Sci World J 2016;2016:7370524.

[4] Daw MA, El Bouzedi A, Ahmed MO, Dau AA, Agnan MM. Epidemiology of hepatitis C virus and genotype distribution in immigrants crossing to Europe from north and sub Saharan Africa. Trav Med Infect Dis 2016:15.

[5] Daw MA, El Bouzedi A. Prevalence of hepatitis B and hepatitis C infection in Libya: results from a national population based survey. BMC Infect Dis January 9, 2014;14:17.

Chapter 3.3

Hepatitis C Virus in Sub-Saharan Africa

Osi Obadahn[1], Sanaa M. Kamal[2]

[1]University of Lagos, Lagos, Nigeria; [2]Ain Shams Faculty of Medicine, Cairo, Egypt

Chapter Outline

INTRODUCTION

Sub-Saharan Africa constitutes the countries located south of the Great Sahara. Sub-Saharan Africa consists of 49 African nations where almost 950 million people live (Fig. 3.3.1). The countries of this region suffer from war, economic cases, poverty, political instability, disease, and hunger. Various viral, bacterial, and parasitic diseases are endemic in Africa [1]. Recently, waves of illegal immigration toward the North represent major health concerns with spread of endemic infections to North Africa and Europe [2].

FIGURE 3.3.1   Sub-Saharan African countries.

HEPATITIS C VIRUS PREVALENCE IN CENTRAL AFRICA

The prevalence for the central African region is about 6.0%. The highest hepatitis C virus (HCV) prevalence is in Cameroon with a prevalence of 13.8%. The lowest prevalence is detected in Equatorial Guinea (1.7%) (Table 3.3.1) [3,4].

Hepatitis C in Gabon

In their meta-analysis of studies performed on HCV patients in Africa, Riou et al. [5] reported an HCV seroprevalence of 4.9% in Gabon, the highest HCV prevalence rate in the Middle African Region [6]. A study identified the risk of transfusion-transmitted viral infections in Gabon by analyzing the records of the Gabonese National Blood Transfusion Centre. According to their analysis, the risk of HCV transmission was estimated at 207.94 cases per million donations [7]. It should be noted that this rate was higher than that for HIV (64.7 per million) but lower than hepatitis B virus (HBV; 534.53 per million). Hepatitis C genotype 4 is the prevalent genotype in Gabon [8,9].

Hepatitis C Infection in the Democratic Republic of Congo

Several studies assessed the prevalence of HCV in Congolese blood donors, pregnant women, military personnel, individuals with HIV, children, patients with sickle cell disease and hospitalized patients, and commercial sex workers. The pooled prevalence of anti-HCV was 2.9% (95% confidence interval

TABLE 3.3.1 Hepatitis C Virus Prevalence Rates in Sub-Saharan Africa

Country	Population (1000)	Range	Hepatitis C Virus Prevalence (%) Average
Central Africa			
Burundi	6356	4.9–33.3	11.3
Cameroon	14,876	0.0–40.0	13.8
CAR	3717	0.0–6.1	2.4
Chad	7885	2.4–5.8	4.8
Congo	3018	2.5–9.2	
DR Congo	50,948	4.3–6.6	5.5
Equatorial Guinea	457	1.7–1.9	1.7
Gabon	1230	6.5–16.5	9.2
Rwanda	7609	0.9–17.0	4.1
Sudan	31,095	1.5–3.2	2.8
Uganda	23,300	0.0–14.2	6.6
Central Africa total	147,474	0.0–40.0	6.0
West Africa			
Benin	6272	0.0–4.0	1.6
Burkina Faso	11,535	2.2–8.3	4.9
Cote d'Ivoire	16,013	3.3–8.2	3.3
Gambia	1303	2.4–3.0	2.4
Ghana	19,306	0.1–5.4	1.7
Guinea	8154	0.8–8.7	5.5
Mauritania	2665	1.1–1.5	1.1
Niger	10,832	0.0–7.6	1.8
Nigeria	113,862	0.0–5.8	2.1
Senegal	9421	0.0–7.3	2.2
Togo	4405	1.3–6.1	3.9
West Africa total	203,766	0.0–8.7	2.4

Continued

TABLE 3.3.1 Hepatitis C Virus Prevalence Rates in Sub-Saharan Africa—cont'd

Country	Population (1000)	Range	Hepatitis C Virus Prevalence (%) Average
South and East Africa			
Eritrea	3659	0.0–6.0	1.9
Ethiopia	62,908	0.6–3.4	1.9
Kenya	30,669	0.0–1.0	0.9
Madagascar	15,970	1.2–3.3	2.1
Malawi	11,308	0.7–1.0	0.7
Mozambique	18,292	2.1–3.2	2.8
Somalia	8778	0.0–7.0	1.5
South Africa	43,309	0.0–3.5	0.1
Swaziland	925	1.2–1.8	1.5
Tanzania	35,119	0.5–8.6	3.2
Zambia	10,421	0.0–0.3	0.2
Zimbabwe	12,627	0.2–7.7	2.0
SE Africa total	253,986	0.0–8.6	1.6
Africa total	605,225	0.0–40.0	3.0

Adapted with permission from Madhava V, Burgess C, Drucker E. Epidemiology of chronic hepatitis C virus infection in sub-Saharan Africa. Lancet Infect Dis May 2002; 2:293–310.

1.5%–4.3%) [10–13]. The seroprevalences were 16.7% for HAV, 24.6% for HBV, 2.3% for HCV, and 10.4% for HEV, and 26.1% of HBV-positive patients were also infected with HDV [11]. Iatrogenic transmission of HCV in mid-20th century seems responsible for the HCV status of Democratic Republic of Congo (DRC) [12,13]. Blood transfusion was a risk factor for HCV transmission. However, the adoption of blood screening in the blood banks resulted in a sharp reduction in HCV seroprevalence from 11.8% to 2.3% from 2004 to 2012 [10]. Recently, a new HCV genotype 7 was recently identified in patients from DRC. This prototype QC69 virus was shown to be a new lineage distinct from genotypes 1–6 [14]. A near-complete genome sequence of this variant BAK1 was performed to the strain obtained from Congolese patients. Evolutionary analysis indicates that this isolate, BAK1, could be the first reported strain belonging to a new HCV-7b subtype. This new subtype has been incorrectly identified as genotype 2 by the Versant HCV Genotype 2.0 assay (LiPA) [14].

HCV PREVALENCE IN THE WEST AFRICAN REGION

The overall prevalence of HCV in West Africa ranges between 1.1% and 5.5% with a mean of 2.4% [5]. The highest HCV prevalence is reported from Guinea where HCV prevalence reaches 5.5%. A cross-sectional survey of individuals aged ≥50 years in Guinea-Bissau revealed that the prevalence of HCV was 4.4% among women and 5.0% among men. In multivariate analysis, the independent risk factors for HCV infection were age [baseline: 50–59 years; 60–69 years, adjusted odds ratio (AOR): 1.67, 95% CI: 0.91–3.06; ≥70 years, AOR: 3.47, 95% CI: 1.89–6.39] and having bought or sold sexual services (AOR: 3.60, 95% CI: 1.88–6.89). The prevalent HCV was genotype 2 [15].

Hepatitis C in Nigeria

Nigeria is the country with the highest population in West Africa. According to a meta-analysis performed by Riou et al. in 2016, the seroprevalence of HCV in Nigeria is estimated at 3.1%, based on 29 studies [5]. The sample sizes of the included studies ranged between 96 and 33,379 patients. A large population survey performed in Nigeria during the period from 2010 to 2012, which screened 5558 Nigerian adults for HCV antibodies, reported a HCV prevalence rate of 0.9%. However, two studies conducted in 2015 showed a seroprevalence of 5.8% [16,17].

HIV/HCV CO-INFECTION IN NIGERIA

HIV/HCV co-infection is a serious problem in Nigeria. A study screened newly infected HIV patients in Ughelli, a suburban area in Nigeria, for HCV antibodies. The seroprevalence was estimated at 15% [18]. In another region in Nigeria, a study performed on 183 HIV-positive patients showed that 16.9% had HIV/HCV/HBV infections simultaneously. HIV/HCV dual infection was reported in 23.5% of the cases [19]. A study on 1239 antenatal women in the Enugu region of Nigeria reported a HCV prevalence rate of 2.6%. The prevalence rate of HIV/HCV co-infection was 0.16% [20]. A study [21] tested 146 HIV patients for HCV antibodies. They observed that 8.2% of the patients had HIV/HCV dual infection. Ezechi et al. reported a HCV prevalence rate of 1.5% among 2391 HIV-positive pregnant women. In addition, 0.08% had HIV/HBV/HCV triple infection [22].

HCV GENOTYPES IN NIGERIA

Forbi et al. [23] studied HCV infections in two communities in North-Central Nigeria and reported a prevalence of 15%. Among the 60 HCV patients, 85% were infected with genotype 1 and 15% were infected with genotype 2. A similar result was reported by Agwale et al. [21] in his study on patients with HIV/HCV coinfection. Out of the 12 patients, 9 had HCV genotype 1 and the remaining 3 had genotype 2. Oni et al., [24] in the study performed in 1996 on 200 blood donors, had a slightly different observation. The authors concluded that genotypes 1 and 4 were prevalent in Nigeria.

MODES OF HCV TRANSMISSION IN NIGERIA

According to a study performed on 420 children in Enugu, Nigeria, the most important risk factor of HCV transmission is traditional scarifications and tattoos. In a study performed on HCV patients having either diabetes or lymphoproliferative disorders, the authors identified having sex with multiple partners as a significant mode of HCV transmission [25]. Opaleye et al. [26] investigated the risk factors for HCV transmission among 182 pregnant women in southwestern Nigeria. However, the authors could not identify any significant correlation between HCV infection and circumcision, tattooing, or incision. Ezechi et al. [22] reported a significant correlation between history of induced abortion and HIV/HCV coinfection.

HCV Infection in Mali

A study showed that HCV seropositivity in Mali is significantly higher in rural areas (7%) than in urban regions (1%). In older persons, Genotypes 1 and 2 were the frequent genotypes in Mali with a predominance of HCV genotype 2. Hospitalization and coinfection with HIV were the major risk factors [27,28]. Another study assessed HIV, HBV, and HCV among blood donors in Mali. The overall seroprevalence of HIV was 0.88%, HBV 5.3%, and HCV 0.55% [29]. In patients undergoing hemodialysis, the prevalence of anti-HCV antibodies was 19.7% [30]. This study determined the prevalence of HIV, HBV, and HCV infections in addition to assessment of the frequency of red cell alloimmunization before and after blood transfusion. The prevalence of viral infections observed at the time of enrolment of patients in the study was 1%, 3%, and 1%, respectively, for HIV, HBV, and HCV. Three cases of seroconversion after blood transfusion were detected, including one for HIV, one for HBV, and one for HCV in sickle cell anemia patients. All these patients had received blood from occasional donors. The red cell alloimmunization was observed in 4.4% of patients.

Hepatitis C in Burkina Faso

Analysis of studies conducted in Burkina Faso showed a HCV prevalence rate of 6.1% [31–34]. They also noted that Burkina Faso had the highest HCV prevalence rate among countries in the West African region. Zeba et al. tested 2200 blood donors for HCV antibodies in 2014 and reported a prevalence rate of 4.4% [31]. In their study on 31,405 first-time blood donors in 2009, Nagalo et al. [32] reported a HCV prevalence rate of 6.3%. The prevalence rate increased to 6.78% among blood donors who have previously donated blood [32,33].

Many HCV patients in Burkina Faso exhibit multiple blood-transmitted infections in concurrence with HCV. Among 4520 blood donors, the seroprevalence of HCV was 8.69%. Of the donors, 3.3% had multiple infections, among them 1.39% had hepatitis B surface antigen (HBsAg)/HCV, whereas 0.11% had HBsAg/HCV/syphilis triple infection [34]. In a study by Zeba et al. [35] on 462 individuals, the prevalence of HCV infection was 3.9%, whereas HCV/HBV coinfection was reported in 2.2%. In a study on 607 pregnant women, 62.27% had HIV infection. In addition, 2.38% of the HIV-positive patients had HCV [29].

HCV Genotypes

According to Zeba et al. [30,34], HCV genotypes 2 and 3 are prevalent in Burkina Faso. Among HCV patients that were identified, 56.3% were infected with genotype 2, whereas 15.6% were infected with genotype 3. According to Simpore et al., the most important genotype in Burkina Faso is 2a, and the prevalence rate may reach 60%. In his study on 547 pregnant women in Ouagadougou, 3.3% had HCV antibodies.

Risk Factors for HCV Transmission in Sub-Saharan Africa

Intravenous drug use is not a major route of HCV transmission in sub-Saharan Africa. Other factors play important role in HCV transmission.

Unsterile Injections

The use unsterile medical injections and instrument within or outside the formal health care system account for significant HCV transmission in Sub-Saharan Africa. A study estimated that Africans get an average of 1.5 medical injections per year of which 50% are unsafe [36]. Another study estimated the risk of HCV transmission through an unsafe injection at 6% [18]. Many studies have found a link with injection use, for example, a group of rural women from Tanzania with a history of hormone injections for contraception had a HCV prevalence rate of 19% compared with 5% for women who used no or other family planning methods [5,35–38].

Stratification and Tattooing

Stratification and tattooing may contribute to the transmission of HCV in Sub–Saharan Africa. However, the relation is hard to be proved [5,7,38].

Heterosexual Transmission

Some studies observed high prevalence of sexually transmitted diseases in Africa including heterosexual transmission of HCV.

Mother-to-Child Transmission

Some of these studies report zero to low HCV prevalence in young children, as evidence that mother-to-child transmission is negligible [39,40]. Studies elsewhere have shown that vertical transmission is proportional to the HCV RNA level in the mother. High viral loads have been demonstrated in women with acute HCV (compared with those with chronic HCV) and in women with concurrent HIV infection. In one study of 441 mother–child pairs, the rate of vertical transmission was 3.8 times higher in those mothers co-infected with HIV [39–41]. No transmission of HCV by breastfeeding has been noted.

HCV and HIV Co-infection

HCV and HIV share a common route of parenteral transmission. Several studies found virtually no association between the prevalence of the two infections. Only one study showed a significant level of co-infection [37].

REFERENCES

[1] The world bank report on Sub-Saharan Africa [Internet]. Available from: http://data.world bank.org/region/sub-saharan-africa.

[2] Gonzalez-Garcia J, Hitaj E, Mlachila M, Viseth A, Yenice M. Sub-Saharan African migration. Washington, DC: International Monetary Fund; November 2016.

[3] Cancre N, Gresenguet G, Mbopi-Keou FX, et al. Hepatitis C virus RNA viraemia in central Africa. Emerg Infect Dis 1999;5:484–5.

[4] Njouom R, Frost E, Deslandes S, Mamadou-Yaya F, Labbé AC, Pouillot R, Mbélesso P, Mbadingai S, Rousset D, Pépin J. Predominance of hepatitis C virus genotype 4 infection and rapid transmission between 1935 and 1965 in the Central African Republic. J Gen Virol October 2009;90(Pt. 10):2452–6.

[5] Riou J, Ahmed M, Blake A, Vozlinsky S, Brichler S, Eholie S, Boelle P, Fontanat A. Hepatitis C virus seroprevalence in adults in Africa: a systematic review and meta-analysis. J Viral Hepat 2016;23:244–55.

[6] Njouom R, Caron M, Besson G, Ndong-Atome GR, Makuwa M, Pouillot R, Nkoghé D, Leroy E, Kazanji M. Phylogeography, risk factors and genetic history of hepatitis C virus in Gabon, Central Africa. PLoS One 2012;7(8):e42002.

[7] Rerambiah LK, Rerambiah LE, Bengone C, Djoba Siawaya JF. The risk of transfusion-transmitted viral infections at the Gabonese national blood transfusion centre. Blood Transfus July 2014;12(3):330–3.

[8] Kamal SM, Nasser IA. Hepatitis C genotype 4: what we know and what we don't yet know. Hepatology April 2008;47(4):1371–83. http://dx.doi.org/10.1002/hep.22127.

[9] Ndong-Atome GR, Makuwa M, Ouwe-Missi-Oukem-Boyer O, Pybus OG, Branger M, Le Hello S, Boye-Cheik SB, Brun-Vezinet F, Kazanji M, Roques P, Bisser S. High prevalence of hepatitis C virus infection and predominance of genotype 4 in rural Gabon. J Med Virol September 2008;80(9):1581–7.

[10] Muzembo BA, Akita T, Matsuoka T, Tanaka J. Systematic review and meta-analysis of hepatitis C virus infection in the Democratic Republic of Congo. Public Health October 2016;139:13–21.

[11] Makiala-Mandanda S, Le Gal F, Ngwaka-Matsung N, Ahuka-Mundeke S, Onanga R, Bivigou-Mboumba B, Pukuta-Simbu E, Gerber A, Abbate JL, Mwamba D, Berthet N, Leroy EM, Muyembe-Tamfum JJ, Becquart P. High prevalence and diversity of hepatitis viruses in suspected cases of yellow fever in the Democratic Republic of Congo. J Clin Microbiol May 2017;55(5):1299–312.

[12] Iles JC, Abby Harrison GL, Lyons S, Djoko CF, Tamoufe U, Lebreton M, Schneider BS, Fair JN, Tshala FM, Kayembe PK, Muyembe JJ, Edidi-Basepeo S, Wolfe ND, Klenerman P, Simmonds P, Pybus OG. Hepatitis C virus infections in the Democratic Republic of Congo exhibit a cohort effect. Infect Genet Evol October 2013;19:386–94.

[13] Batina Agasa S, Dupont E, Kayembe T, Molima P, Malengela R, Kabemba S, Andrien M, Lambermont M, Cotton F, Vertongen F, Gulbis B. Multiple transfusions for sickle cell disease in the Democratic Republic of Congo: the importance of the hepatitis C virus. Transfus Clin Biol October 2010;17(4):254–9.

[14] Salmona M, Caporossi A, Simmonds P, Thélu MA, Fusillier K, Mercier-Delarue S, De Castro N, LeGoff J, Chaix ML, François O, Simon F, Morand P, Larrat S, Maylin S. First next-generation sequencing full-genome characterization of a hepatitis C virus genotype 7 divergent subtype. Clin Microbiol Infect November 2016;22(11):e1-947.

[15] Plamondon M, Labbé AC, Frost E, Deslandes S, Alves AC, Bastien N, Pepin J. Hepatitis C virus infection in Guinea-Bissau: a sexually transmitted genotype 2 with parenteral amplification? PLoS One April 18, 2007;2(4):372.

[16] Onyekwere C, Hameed L. Hepatitis B and C virus prevalence and association with demographics: report of population screening in Nigeria. Trop Doct 2015;45(4):231–5.

[17] Owolabi O, Adesina K, Fadeyi A, Popoola G. Hepatitis C Virus sero-prevalence, antigenaemia and associated risk factors among pregnant women in Nigeria, Ethiop. Med J 2015;53(4):173–81.

[18] Newton OE, Oghene OA, Okonko IO. Anti-HCV antibody among newly diagnosed HIV patients in Ughelli, a suburban area of Delta State Nigeria. Afr Health Sci 2015;15(3): 728–36.

[19] Ogwu-richard SO, Ojo DA, Akingbade OA. Triple positivity of HBsAg, anti-HCV antibody, and HIV and their influence on CD^{4+} lymphocyte levels in the highly HIV infected population of Abeokuta, Nigeria. Afr Health Sci 2015;15(3):11–3.

[20] keako, Ezegwui H, Ajah L, Dim C, Okeke T. Seroprevalence of human immunodeficiency virus, hepatitis B, hepatitis C, syphilis, and Co-infections among antenatal women in a tertiary institution in south east, Nigeria. Ann Med Health Sci Res 2014;4(6):954–8.

[21] Agwale S, Tanimot L, Womac K, Odama L, Leung K, Duey D, Negedu R, Momoh I, Audu S, Mohammed U, Inyang B, Graham B, Ziermann R. Prevalence of HCV coinfection in HIV-infected individuals in Nigeria and characterization of HCV genotypes. J Clin Virol 2004;31:S3–6.

[22] Ezechi OC, Kalejaiye OO, Gab-Okafor CV, Oladele DA, Oke BO, Musa ZA, Ekama SO, Ohwodo H, Agahowa E, Gbajabiamilla T, Ezeobi PM, Okwuraiwe A, Audu R RA, Okoye RN, David AN, Odunukwe NN, Onwujekwe DI, Ujah IA. Sero-prevalence and factors associated with hepatitis B and C co-infection in pregnant Nigerian women living with HIV infection. Pan Afr Med J March 13, 2014;17:197.

[23] Eke C, Ogbodo S, Ukoha O, Muoneke V, Ibekwe R, Ikefuna A. Seroprevalence and correlates of hepatitis C virus infection in secondary school children in Enugu, Nigeria. Ann Med Health Sci Res 2016;6(3):156–61.

[24] Onyekwere CA, Ogbera AO, Dada AO, Adeleye OO, Dosunmu AO, Akinbami AA, Osikomaiya B, Hameed O. Hepatitis C virus (HCV) prevalence in special populations and associated risk Factors : a report from a tertiary hospital. Hepat Mon 2016;16(5):1–5.

[25] Opaleye O, Igboama M, Ojo J, Odewale G. Seroprevalence of HIV, HBV, HCV, and HTLV among pregnant women in southwestern Nigeria. J Immunoassay 2016;37(1):29–42.

[26] Koné MC, Sidibé ET, Mallé KK, Beye SA, Lurton G, Dao S, Diarra MT, Dao S. Seroprevalence of human immunodeficiency virus, hepatitis B virus and hepatitis C virus among blood donors in Segou, Mali. Med Sante Trop January–March 2012;22(1):97–8.

[27] Bouare N, Gothot A, Delwaide J, Bontems S, Vaira D, Seidel L, Gerard P, Gerard C. Epidemiological profiles of human immunodeficiency virus and hepatitis C virus infections in Malian women: risk factors and relevance of disparities. World J Hepatol April 27, 2013;5(4):196–205. http://dx.doi.org/10.4254/wjhv5i4.196.

[28] Bouare N, Vaira D, Gothot A, Delwaide J, Bontems S, Seidel L, Gerard P, Gerard C. Prevalence of HIV and HCV infections in two populations of Malian women and serological assays performances. World J Hepatol December 27, 2012;4(12):365–73.

[29] Baby M, Fongoro S, Konaté MK, Diarra A, Kouriba B, Maïga MK. Prevalence and risk factors of hepatitis C virus infection in chronic hemodialysis patients at the University Hospital of Point G, Bamako, Mali. Mali Med 2011;26(2):12–5.

[30] Zeba M, Sanou M, Bisseye C, Kiba A, Marius B, Djigma FW, Compaoré TRY, Nebié YK, Kienou K, Sagna T, Pietra V, Moret R, Simporé J. Characterisation of hepatitis C virus genotype among blood donors at the regional blood transfusion centre of Ouagadougou, Burkina Faso. Blood Transfus 2014;12:54–7.

[31] Nagalo BM, Bisseye C, Sanou M, Kienou K, Nebie YK, Kiba A, Dahourou H, Ouattara S, Nikiema J, Moret R, Zongo J, Simpore J. Seroprevalence and incidence of transfusion-transmitted infectious diseases among blood donors from regional blood transfusion centres in Burkina Faso, West Africa. Trop Med Int Heal 2012;17(2):247–53.

[32] Nagalo MB, Sanou M, Bisseye C, Kaboré MI, Nebie K, Kienou K, Kiba A, Dahourou H, Ouattara S, Didier J, Simporé J. Seroprevalence of human immunodeficiency virus, hepatitis B and C viruses and syphilis among blood donors in Koudougou (Burkina Faso) in 2009. Blood Transfus 2011;9(4):419–24.

[33] Zeba M, Ouattara C, Karou S, Bisseye C, Ouermi D, Djigma F, Sagna T, Pietra V, Moret R, Nikiema J, Simpore J. Prevalence of HBV and HCV markers among patients attending the Saint Camille medical centre in Ouagadougou. Pakistan J Biol Sci 2012;15(10):484–9.

[34] Zeba MT, Karou SD, Sagna T, Djigma F, Bisseye C, Ouermi D, Pietra V, Pignatelli S, Gnoula C, Sia JD, Moret R, Nikiema J, Simpore J. HCV prevalence and co-infection with HIV among pregnant women in Saint Camille Medical Centre, Ouagadougou. Trop Med Int Heal 2011;16(11):1392–6.

[35] Madhava V, Burgess C, Drucker E. Epidemiology of chronic hepatitis C virus infection in sub-Saharan Africa. Lancet Infect Dis May 2002;2:293–310.

[36] Barth RE, Huijgen Q, Taljaard J, Hoepelman AI. Hepatitis B/C and HIV in sub-Saharan Africa: an association between highly prevalent infectious diseases. A systematic review and meta-analysis. Int J Infect Dis 2010;14:e1024–31.

[37] Nicot T, Rogez S, Denis F. Epidemologie de l'hepatite C en Afrique. Gastroenterol Clin Biol 1997;21:596–606.

[38] Menendez C, Sanchez-Tapias JM, Kahigwa E. Prevalence and mother-to-infant transmission of hepatitis viruses B, C, and E in southern Tanzania. J Med Virol 1999;58:215–20.

[39] Opaleye OO, Igboama MC, Ojo JA, Odewale G. Seroprevalence of HIV, HBV, HCV, and HTLV among pregnant women in Southwestern Nigeria. J Immunoassay Immunochem 2016;37(1):29–42.

[40] Tajiri H, Miyoshi Y, Funada S, et al. Prospective study of mother-to-infant transmission of hepatitis C virus. Paediatr Infect Dis J 2001;20:10–4.

[41] Develoux M, Meynard D, Delaporte E. Low rate of hepatitis C virus antibodies in blood donors and pregnant women from Niger. Trans R Soc Trop Med Hyg 1992;86:553.

Chapter 3.4

Hepatitis C Virus Infection in the Indian Sub-Continent

Shashi Shekhar

Dar Al Uloom University College of Medicine, Riyadh, Kingdom of Saudi Arabia

Chapter Outline

Hepatitis C in Developing Countries. https://doi.org/10.1016/B978-0-12-803233-6.00008-4

83

INTRODUCTION

Natural History of Hepatitis C Virus

Hepatitis C virus (HCV) first identified in 1989 produces a slowly progressive liver disease, namely hepatitis, cirrhosis, and hepatocellular carcinoma (HCC). It is affecting 170 million people (3%) world over [1]. More than 3 million people are affected annually (Fig. 3.4.1).

The genome of HCV is a single-stranded, positive-sense RNA molecule (++ss RNA) of approximately 9.6 kb in length [2]. The HCV antigen is of core variety.

The incubation period is 50–150 days. HCV spreads through blood, sexual activity when mixed with infected blood, and through the placenta.

No vaccine is available against HCV.

Primary Causes of Hepatitis

Primary causes of hepatitis are HCV, HBV infection, and alcoholism.

Primary Causes of Chronic Liver Disease (CDC)

1. Hepatitis C virus infection	26%
2. Alcohol	24%
3. Hepatitis C virus + alcohol	14%
4. Unknown	17%
5. Hepatitis B virus infection	11%
6. Hepatitis B virus + alcohol	03%
7. Other	15%

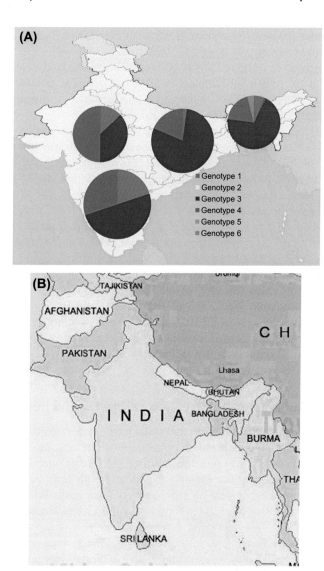

FIGURE 3.4.1 (A) Hepatitis C virus genotypes in regions of India. (B) Hepatitis C virus incidence and genotypes in nine countries of the Indian subcontinent. (1) Pakistan 6%–6.8%, genotypes 3a, 2a; (2) Myanmar 0.34%–2.3%; (3) Bhutan 1.3%; (4) Afghanistan 1%; (5) Bangladesh 0.6%; (6) Nepal 0.6%; (7) India 0.33%, range 0.5%–1.5%, genotypes 1,3,4,6; (8) Sri Lanka 0.16%; (9) Maldives—not known.

Hepatitis C Infection in the Indian Subcontinent

The Indian subcontinent is situated between the foothills of Himalayas and shores of the Indian Ocean. This area is known by various names, i.e., South Asia, Southern Asia, and SAARC. It is the conglomeration of nine countries, namely, India, Pakistan, Myanmar, Afghanistan, Nepal, Bangladesh, Sri Lanka, Bhutan, and Maldives.

The land area of world is 134,940,000 km^2, whereas the land area of the nine countries in the Indian subcontinent is 5,809,045 km^2. The population of the world (2016) is 7,465,343,800, whereas the population of the nine countries in the Indian subcontinents (2016) is 1,793,307,209.

The incidence of HCV world over is 2.2% [3], but the incidence of HCV in the Indian subcontinent varies in the nine countries. In Pakistan it is 6%–6.8%, the second highest in the world; in Bhutan 1.3%; in Myanmar 0.34%–2.03%; in Afghanistan 1.0%; in Nepal 0.6%; in Bangladesh 0.6%; in India 0.33%; and in Sri Lanka 0.16%; the incidence in Maldives is not known.

The studies conducted on HCV virus genotype world over (study of Messina et al. [3]), which included 1217 publications from 117 countries covering 90% of the global population, concluded that HCV genotype 1 is the most prevalent worldwide, comprising 83.4 million cases (46.2% of all HCV cases); HCV genotype 3 is the most common globally comprising 54.3 million cases (30.15% of all HCV cases); HCV genotypes 2, 4, and 6 are responsible for 22.8% of HCV cases; and HCV genotype 5 is prevalent in less than 1% of all HCV cases. The HCV genotype distribution in the Indian subcontinent is as follows; in India Type 3 (63.85%), Type 1(25.72%), and Type 4 (7.50%); in Pakistan, it is Type 3 (61.3%) and Type 2 (11.3%–17.3%).

Safe Injection Practice Intervention in Pakistan and India

Available data show that Pakistan and India have taken steps toward achieving safe injection practice through potential interventions, i.e., introduction of syringe and needle reuse prevention device [4]. Use of injections in health care is very common in the Indian subcontinent with rates being 2.4–13.6 injections per person per year on cross-country level in the Indian subcontinent, though information is limited for some countries, i.e., Bhutan, Maldives, and Sri Lanka.

In the Indian subcontinent, injections with reused syringe are reported in 5%–50% cases, which may be a potential source of blood-borne pathogen transmission. Injection is given by private qualified and nonqualified practitioners and is the preferred method of treatment among the rural uneducated poor in some countries [4].

The Treatment Options for HCV

Interferon: The treatment of HCV started with "Interferon," which requires more than 70 painful injections that have a high failure rate and heavy side effects, which makes the patients too weak to go to work [5].

Sofosbuvir: New molecule is a much improved treatment for hepatitis C infection, which cures HCV in 12 weeks and needs one tablet a day for 12 weeks. Sofosbuvir is a direct-acting antiviral capable of revolutionizing the treatment. The cure rates reported are as high as 90% with some HCV genotypes.

Gilead Sciences (USA) was the first manufacturer of Sofosbuvir and priced it as USD 1000 per tab, USD 84,000 for a 12-week course [5].

Innovative Financing Model for Free Treatment of HCV in India and Pakistan

In India "Jeevan Rekha scheme in Haryana and Punjab province" (Shimona Kanwar TOI) [15] provides free treatment by Sofosbuvir tablets against HCV infection for the weaker section of the society of BPL/SC (Below Poverty Line/Scheduled Caste) category. They screened 18,000 persons for HCV and HBV and provided free treatment to 1100 patients and treatment at subsidized rates to 500 patients.

In Pakistan "Government has fixed the price of Sofosbuvir tablet at Rs. 5868 for 28 tablets [6].

Public Education Initiatives Against HCV

Clinicians, public health professionals, NGOs, patient support group, enthusiasts, advocates, and industry have started playing their role in public education against the spread of HCV and regarding treatment initiatives in India [7]. The *Sixth World Hepatitis Day – July 28, 2016* was organized by the Government of India to focus on HCV infection.

HCV IN INDIA

Population 1,308,221,797; Land Area 1,308,221,797; per Capita Income USD 1688

Incidence of HCV 0.33% (0.5%–1.5%)

HCV infection is a major public health problem worldwide. Overall, 12 million people in India are infected with HCV, for which no vaccine is yet available.

Over the past several decades, the incidence of HCV is on the rise in India. Though 3% of the world's population is infected with HCV, the overall HCV infection rate in India is only 0.33%. In terms of endemicity, India comes in low or intermediate zone. India having the second largest population in the world has a huge variation of epidemiology among its 35 states. The incidence of HCV has nationwide variation, with 1.85% in North India (Delhi); 36.9% in South India (Chennai); 33.8% in Central India (Punjab); and 74% in North-eastern India (Manipur) [8].

Co-infection HCV and HIV

HCV and HIV co-infection is rare in India. Drug-induced toxicity may be more common among HIV/HCV co-infection, particularly with the use of HIV-1 protease inhibitors and anti-tubercular drugs.

Modes of Transmission of HCV in India

Blood and blood products: HCV is acquired through transfusion of blood and blood products in India. India has introduced mandatory policy of HCV testing in all blood banks throughout the country. In seven Indian cities, including Delhi and Mumbai , blood samples were studied from eight blood banks for sero-prevalence of HCV (by anti-HCV antibodies) and the result was positive in 0.33%.

Occupational exposure: After accidental percutaneous exposure from HCV-positive source, anti-HCV seroconversion rate is 1.8%.

HCC and HCV in India

Liver cancer ranks fifth among the most frequent cancer in both genders in Cancer Registries of the Indian population based on five urban cities of Mumbai, Bangalore, Chennai, Delhi, and Bhopal. The main cause of HCC in India is alcohol consumption and HBV and HCV infection. HCC-positive marker for HCV was present in 12.21% and signal-to-cut off (s/co) ratio was 5.45% and heavy alcohol intake was found in 15.96% and signal-to-cut off (s/co) ratio was 2.83%. A synergistic effect between HCV/ALCOHOL was found with a synergy index 1.257.

Hemodialysis and HCV in India

In a tertiary care hospital in Faridkot, Punjab, 33.8% patients on hemodialysis were having HCV infection, and HCV/HBV co-infection was found in 0.8%.

High burden of HCV/HIV co-infection in people who inject drugs in Northeast India (Manipur).

HCV prevalence in one north-eastern Indian State Manipur is 74%. HCV/ HIV co-infection rate is very high as of 29% and these are people who inject drugs (PWID). Among 31% of HIV-positive PWID, 95% were co-infected with HCV. HCV infection was found associated with two districts namely Churachandpur and Bishnupur of Manipur, longer duration of injecting drugs, Injecting Drugs at least once daily, generally reuse of needle and syringe for injecting drug and lastly HIV positivity. This needs urgent and effective prevention, diagnosis, and treatment.

HCV and HIV Burden in PWID in South India (Chennai) [9]

CHHERS study in Chennai observed death caused by liver disease progression in autopsy findings of HCV- and HIV-infected dead patients. Among 36.9%

HCV-infected and 16.7% of HIV-infected patients, 71.6% had mild/no stiffness, 14.9% had moderate stiffness, and 13.55 had severe stiffness or cirrhosis. Among PWID cases mortality rate ratio (MMR) was significantly higher in those who had moderate liver stiffness (MMR = 2.31) and severe stiffness (MMR = 4.86).

HCV and HIV Burden in North India (Delhi)

In one study [10], the prevalence of HCV seropositivity was less in HIV-positive patients between 2012 and 2014 in a superspecialty hospital in Delhi. Only one patient was found to be anti-HCV antibody positive in 3 years period.

In a study conducted in AIIMS, Delhi [7], anti HCV antibody was found present in 1.85% of healthy blood donors, and in 13.83% of donors with history of acute hepatitis or chronic liver disease.

Lab Diagnosis of HCV in India

Long window period in HCV infection over 60 days leads to slow development of antibody marker and rapid viral multiplication, with a doubling time of 0.45 days causing high viremia within a short period of few days. This leads to high probability of HCV transmission through blood component transfusion or through shared needle in intravenous drug users and in unsafe sexual practitioners. In the Indian context, the specificity of HCV antibody detection in serology tests is low. Thus, specific supplemental assays, i.e., RIBA/NAT, need to be used for confirmation.

Routine lab diagnosis of HCV in India is based on detecting specific antibodies by enzyme immunoassay (EIA) or by chemiluminescence immunoassay (CIIA) and by signal-to-cut off (s/co) ratio in anti-HCV antibody tests, i.e., EIA or by CIIA. An s/co ratio of ≥6 was 95% sensitive and 92.5% specific.

Treatment of HCV in India

Standard of care of chronic hepatitis C in India is "Peg interferon" and "Ribavirin," though the response to treatment is unclear. Sofosbuvir, the new molecule, is a much improved treatment for hepatitis C infection, which cures HCV in 12 weeks and needs one tablet a day for 12 weeks. Sofosbuvir is a direct-acting antiviral capable of revolutionizing the treatment. The cure rates reported are as high as 90% with some HCV genotypes.

Sixth World Hepatitis Day – July 28, 2016

WHO established World Hepatitis Day on July 28, 2010, to increase awareness and understanding of viral hepatitis. An estimated 1.4 million persons die each year from some form of hepatitis.

HCV IN PAKISTAN

Population 194,881,548; Land Area 803,940 km^2; per Capita Income USD 1427.08

Incidence of HCV 6%–6.8% (rural 25%, transfusion 2.45%)

With the second highest prevalence of HCV ranging from 4.8% to 8%, the disease of HCV is a major health problem in Pakistan. The incidence is as high as 40% among high-risk group, which includes blood donors, health professionals, drug abusers, and chronic liver disease patients [11].

With up to 5% of the population infected with HCV, Pakistan has the second highest prevalence. It is a big problem in cities like Karachi, where up to one million people are potentially infected with HCV.

In a systemic review of data published between 2010 and 2015, HCV positivity was detected in 6.8% of Pakistani general adult population and active HCV infection was noted in 6% of population [12]. Among blood donors, HCV seroprevalence was 2.45% [12]. In non-liver disease–related complaints, HCV positivity was 61.3%, indicating an increase in nosocomial HCV spread [12].

HCV is endemic in Pakistan and the burden is expected to increase mainly because of widespread use of unsafe medical procedures [12].

Medecins Sans Frontiers – MSF Initiative in Pakistan

"Medecins Sans Frontiers-MSF" since 2015 is running a HCV clinic in Machar colony slum for HCV testing and free treatment of around 400 patients [5].

Government of Pakistan Fixed the Price of "Sofosbuvir"

Government of Pakistan fixed the price of Sofosbuvir tablet at Rs. 5868 for 28 tablets and issued registration letters to 9 local manufacturers and two importing firms for enhancing production of this direct-acting antiviral drug [6].

HCV IN MYANMAR

Population 54,363,426; Land Area 676,577 km^2; per Capita Income USD 824.19

Incidence of HCV 0.34%–2.3%; Disease Burden 1.30 million

HCV is an emerging health problem in Myanmar. Around 1.9 million people are estimated to be infected with HCV [13]. Among blood donors, the prevalence of hepatitis C varied between 0.34% and 2.03% in a study conducted in 2005–7 [14].

In Yangon blood banks in 2000–3 anti-HCV positivity rate was 2.8% [20]. In HCC cases, HCV infection was found in 33% [20].

HCV IN AFGHANISTAN

Population 25,500,100 [16]; Land Area 645,807 km^2; per Capita Income USD 600

Incidence of HCV 1%–1.1%

Afghanistan remains at early epidemic phase of HCV with incidence appearing to be around 1%. Among PWID, incidence is expected to be high. As a result of more than 20 years of political and wartime unrest, the worrying security situation in Afghanistan forbids a national population-based survey for HCV. Hence, no data are available for high-risk population i.e., people undergoing hemodialysis, people suffering from thalassemia, people suffering from hemophilia, PWID, and prisoners [15].

In one publication [16], HCV prevalence in Afghanistan was found in 1.1% of population. People at high risk for HCV included PWID who share needles, female sex workers, truck drivers, prisoners, and homosexual men. Among obstetric population, HCV prevalence is 0.3% [16]. A study showed that 1.92% of sex workers are HCV positive [16]. Among blood donors from general population, HCV positivity was 1.9% [16].

HCV IN NEPAL

Population 26,494,504; Land Area 147,181 km^2; per Capita Income USD 734

Incidence of HCV 0.6%

One study from Nepal [17] reports HCV positivity in 0.6% of healthy adults. Among viral hepatitis, HCV infection was reported in 1.3%. Among intravenous drug abusers, 94% were positive for HCV.

HCV IN BANGLADESH

Population 161,468,818; Land Area 143,998 km^2; per Capita Income USD 1284

Incidence of HCV 0.6%

The prevalence of HCV in Bangladesh is reported to be 0.6% among adult rural population. In a small study of 417 people between 2010 and 2011 at BSMM University, Dhaka, HCV genotype 3 was found in 50.19%, combination of genotypes 3 and 4 in 28.77%, and genotype 1 in 14.14% [18].

TABLE 3.4.1 Hepatitis C in the Indian Subcontinent

HCV in Indian Subcontinent; World Hepatitis Day July 28 (2010 First)

	Land Area (km²)	Population	Per Capita Income	HCV Incidence (%)	Disease Burden	Genotype
World (2016) Total population of 189 economies (2016)	134,940,000	7,465,343,800 (7.292 billions)	GDP of world $10,313	2.2%	185 million	1 (46.2%)83.4 million 3 (30.1%)54.3 million 2,4,6 (22.8%) 5(<1%)
1. India (2016)	3,287,240	1,308,221,797	Dollar 1688	0.33% (0.5%–1.5%)	12 million	3 (71.6%); 1 (23.2%); 4 (2.5%); 2 (2.4%); 5 (0.1%); 6 (0.1%) 1,3 most common; 4,5 in Lower income
		• North – Punjab		33.3% on Hemodialysis; 13.83% in Chronic liver disease		3 (63.85%); 6 (02.70%); 1 (25.72%); 4 (07.50%)
		• North – Delhi		1.8% in Blood donor		
		• NE – Manipur		74.0%		
		• South – Chennai		36.9%		
2. Pakistan (2016)	803,940	194,881,548	1427.08	6%–6.8%	Rural 25%; Blood Transfusion 2.45%	3a (61.3%), 2a (11.3%–17.3%)
3. Myanmar (2014)	676,577	54,363,426	824.19 (2011)	0.34%–2.3%	1.30 million	
4. Afghanistan (2013)	645,807	25,500,100	0.600	1.0%		
5. Nepal (2011)	147,181	26,494,504	734	0.6%		
6. Bangladesh (2016)	143,998	161,468,818	1284	0.6%		
7. Sri Lanka (2016)	65,610	21,203,000	3818	0.16% of Blood donors		
8. Bhutan (2016)	38,394	780,516	2836	1.3%		
9. Maldives	00,298	393,500	9126	Not Known		
Total	**5,809,045**	**1,793,307,209**				

HCV IN SRI LANKA

Population 21,203,000; Land Area 65,610 km^2; per Capita Income USD 3818

Incidence of HCV 1.06% of blood donors

The incidence of HCV in blood donors in Sri Lanka is low, around 1.06%. The incidence among those with alcoholic cirrhosis is around 14.95% [19]. All HCV viruses belonged to genotype 3 [2,14,17,19–21].

HCV IN BHUTAN

Population 780,516; Land Area 38,394 km^2; per Capita Income USD 2836

Incidence of HCV 1.3%

A study conducted on 1666 healthy persons and 440 pregnant women for screening for HCV reported (Da Villa G et al., 1997) low prevalence as anti-HCV positivity was noted only in 1.3% of subjects [21] (Table 3.4.1).

HCV IN MALDIVES

Population 393,500; Land Area 298 km^2; per Capita Income USD 9126

Incidence of HCV – Not Known

Maldives, with no reported HCV infection rate, has no national policy related to screening for HCV. But people testing positive for HCV are registered by name in a confidential system. The testing for HCV is free for all, but compulsory for foreign nationals, pregnant mothers, and presurgical patients. Public-funded treatment for HCV is available for the entire population [20].

REFERENCES

[1] Tiwari AK, Pandey PK, Negi A, Bagga R, Shanker A, Baveja U, Vimarsh R, Bhargava R, Dara RC, Rawat G. Establishing a sample-to cut-off ratio for lab-diagnosis of hepatitis C virus in Indian context. Asian J Transfus Sci July–December 2015;9(2):185–8.

[2] Senevirathna D, Amuduwage S, Weerasingam S, Jayasinghe S, Fernandopulle N. Hepatitis C virus in healthy blood donors in Sri Lanka. Asian J Transfus Sci January 2011;5(1):23–5.

[3] Messina JP, Humphreys I, Flaxman A, Brown A, Cooke GS, Pybus OG, Barnes E. Global distribution and prevalence of hepatitis C virus genotypes; J_ID: HEP, Ref. No. 14–0705.R1. July 19, 2014. p. 1.

[4] Janjua NZ, Ahmad Butt Z, Mahmood B, Altaf A. Towards safe injection practices for prevention of hepatitis C transmission in South Asia: challenges and progress. World J Gastroenterol July 7, 2016;22(25):5837–52. www.wjgnet.com.

[5] Medecins Sans Frontier. Battling hepatitis C and fear in Pakistan, *Voices from the field*, September 14, 2016, MSF USA 333 Seventh Avenue, New York 10001-5004|212-679-6800.

[6] The News–The price of the drug has been fixed at Rs 5868 (28 tablets) Pak.

[7] Kanwar S. Hepatitis C infection spreading fast in India. Chandigarh, India: The Times of India; April 28, 2015. TNN.

[8] Kermode M, Nuken A, Mahanta J. High burden of hepatitis C & HIV co-infection among people who inject drugs in Manipur, Northeast India. Indian J Med Res March 2016;143(3):348–56.

[9] Mehta SH, McFall AM, Srikrishnan AK, Suresh Kumar M, Nandagopal P, Cepeda J, Thomas DL, Sulkowski MS, Solomon SS. Morbidity and mortality among community-based people who inject drugs with a high hepatitis C and human immunodeficiency virus burden in Chennai, India. Open Forum Infect Dis September 2016;3(3):ofw121.

[10] Sharma A, Halim J, Loomba PS. Time trends of seroepidemiology of hepatitis C virus and hepatitis B virus coinfection in human immunodeficiency virus-infected patients in a Super Specialty Hospital in New Delhi, India: 2012–2014. Indian J Sex Transm Dis January–June 2016;37(1):33–7.

[11] Jiwani1 N, Gul Dr R. A silent storm: hepatitis C in Pakistan. JIPM October–December 2011; 1(3).

[12] Umer M, Iqbal M. Hepatitis C virus prevalence and genotype distribution in Pakistan: comprehensive review of recent data. World J Gastroenterol January 28, 2016;22(4):1684–700.

[13] Nearly 10% of Myanmar's population has hepatitis: survey by Coconuts Yangon. January 12, 2016: 11:33 MMT.

[14] Myo-Khina, San-San-Ooa, May Ooa K, Shimonob K, Norio Koidec, Okadac S. Prevalence and factors associated with hepatitis C virus infection among Myanmar blood donors. Acta Med Okayama 2010;64(5):317–21.

[15] Chemaitelly H, Mahmud S, Rahmani AM, Abu-Raddad LJ. The epidemiology of hepatitis C virus in Afghanistan: systematic review and meta-analysis. Int J Infect Dis 2015;40.

[16] Khan S, Attaullah S. Share of Afghanistan populace in hepatitis B and hepatitis C infection's pool: is it worthwhile? Virol J May 11, 2011:20118–216.

[17] Shrestha SM, Subedi NB, Shrestha S, Maharjan KG, Tsuda F, Okamoto H. Epidemiology of hepatitis C virus infection in Nepal. Trop Gastroenterol July–September 1998;19(3):102–4.

[18] Islam MS, Miah MR, Roy PK, Rahman O, Siddique AB, Chowdhury J, Ahmed F, Rahman S, Khan MR. Genotypes of hepatitis C virus infection in Bangladeshi population. Mymensingh Med J January 2015;24(1):143–51.

[19] De Silva HJ, Vitarana T, Ratnatunga N, Breschkin A, Withane N, Kularatne WN. Prevalence of hepatitis C virus markers in Sri Lankan patients with alcoholic cirrhosis. J Gastroenterol Hepatol July–August 1994;9(4):381–4.

[20] WHO South-East Asia Region. Global policy report on the prevention and control of viral hepatitis; 2016.

[21] Da Villa G, Andjaparidze A, Cauletti M, Franco E, Roggendorf M, Sepe A, Zaratti L. Viral hepatitis in the Bhutanese population: preliminary results of a seroepidemiological investigation. Res Virol March–April 1997;148(2):115–7.

FURTHER READING

[1] Bhate P, Saraf N, Parikh P, Ingle M, Phadke A, Sawant P. Cross sectional study of prevalence and risk factors of hepatitis b and hepatitis c infection in a rural village of India. Arq Gastroenterol 2015.

[2] Singh SK, Singh S, Srivastava MK. Co-infection of hepatitis B virus and hepatitis C virus with human immunodeficiency virus infection: a cross-sectional study. Indian J Sex Transm Dis January–June 2016;37(1):95–6.

[3] S. Vasudevan, Shalimar, S.K. Acharya. Demographic profile, host, disease & viral predictive factors of response in patients with chronic hepatitis C virus infection at a tertiary care hospital in north India; Indian J Med Res; [Medknow Publications].

[4] Xu Zhu R, Seto W-K, Lai C-L, Yuen M-F. Gut and liver. Epidemiol Hepatocell Carcinoma Asia Pacific Region May 2016;10(3):332–9.

[5] Malhotra R, Soin D, Grover P, Galhotra S, Khutan H, Kaur N. Hepatitis B virus and hepatitis C virus co-infection in hemodialysis patients: a retrospective study from a tertiary care hospital of North India;. J Nat Sci Biol Med January–June 2016;7(1):72–4.

[6] Kumar R, Gupta S, Kaur A, Gupta M. Individual donor-nucleic acid testing for human immunodeficiency virus-1, hepatitis C virus and hepatitis B virus and its role in blood safety. Asian J Transfus Sci July–December 2015;9(2):199–202.

[7] World Hepatitis Day — July 28, 2015. Additional information about World Hepatitis Day is available at: http://worldhepatitisday.org. Resources for health professionals are available at: http://www.cdc.gov/hepatitis.

[8] Panigrahi AK, Panda SK, Dixit RK, Rao KV, Acharya SK, Dasarathy S, Nanu A. Magnitude of hepatitis C virus infection in India: prevalence in healthy blood donors, acute and chronic liver diseases. J Med Virol March 1997;51(3):167–74.

[9] List of countries and territories by population density From Wikipedia, the free encyclopedia, Population density (people per km^2) by country in 2015.

[10] Kumar T, Shrivastava A, Kumar A, Kayla Laserson F, Jai Narain P, Venkatesh S, Lakhbir S, Chauhan, Averhoff F. Morbidity and mortality weekly report (MMWR). Viral Hepat Surveill India 2011–2013 July 24, 2015;64(28):758–62.

[11] Lokesh U, Srinidhi D, Reddy KS. Post exposure prophylaxis to occupational injuries for general dentist. Indian Prosthodont Soc December 2014;14(Suppl. 1):S1–3.

Chapter 3.5

Hepatitis C in Developing Countries in Southeast Asia

Thi Q. Doan

Hanoi Medical University, Ho Chi Minh City, Vietnam

Chapter Outline

EPIDEMIOLOGY

The epidemiology of hepatitis C virus (HCV) in developing countries in Southeast Asia is not adequately studied. Several observational studies based in Asia reveal an increase in HCV prevalence with age, potentially because of more recent and improved screening of blood products and adoption of safer injection practices [1,2]. National seroepidemiologic studies reveal an overall anti-HCV prevalence between 0.4% and 6%, and smaller studies describe large variances by geographical region [3,4]. The prevalence of HCV in the general population is variable among East Asian countries, ranging from about 0.5% in Singapore and Hong Kong to around 6% in Vietnam and Thailand [5,6] and exceeding 10% in Myanmar [7] The reported prevalence in China is approximately 2%–3%, which amounts to approximately 30 million people [3]. The increasing incidence of hepatocellular carcinoma (HCC) in many countries in recent decades is likely attributable to HCV [3,4] because universal hepatitis B virus (HBV) vaccination has caused dramatic decreases in HBV-related HCC (Table 3.5.1).

GEOGRAPHICAL DISTRIBUTION OF HCV GENOTYPES IN SOUTHEAST ASIA

Hepatitis C genotype 1b is most common in China, Hong Kong, and Taiwan with the genotype 6a found in various parts of the Pearl River Delta of China and Hong Kong [6,7]. Genotypes 1b and 2a are found in Japan and Korea;

genotype 6 is dominant in Cambodia, Laos, Myanmar, and Vietnam and is the dominant variant [7,8] (Figure 3.5.1).

TABLE 3.5.1 Prevalence of Hepatitis C Virus in Southeast Asian Countries

Country	Author	Prevalence (%)
Cambodia	Akkarathamrongsin et al. (2011)	2.3
Indonesia	Sulaiman et al. (1996)	2.1
Laos	Jutavijittum et al. (2007)	1.1
Myanmar	Myo-Khin et al. (2010)	0.95
Philippines	Yanase et al. (2007)	0.4
Singapore	Wang et al., (1995)	0.37
Thailand	Sunanchaikarn et al. (2007)	2.7
Vietnam	WHO (2011)	6.1

HCV GENOTYPE 6 IN SOUTHEAST ASIA

Genotype 6 is highly prevalent in Southeast Asia and shows marked diversity in endemic areas. The HCV genotype 6 strains isolated from East Asia were so divergent that they were initially classified as separate genotypes, designated 7, 8, and 9 [7]. However, further virologic analysis reclassified such strains as individual subtypes within genotype 6 [7]. Genotype 6 infections are also of considerable epidemiological importance: there are an estimated 62 million HCV-infected people in the WHO-defined Western Pacific region, which represents approximately one-third of all infections worldwide [3,4,8]. Genotype 6 and the genotype distribution of HCV infection are variable among and within different East Asian countries. For example, genotype 6 seems to be the most frequent genotype in Myanmar (49% of infections) (29) and Vietnam (52% of infections) (37), but not in Thailand, where the globally distributed subtype 3a, which is associated with injection drug use, is twice as common as genotype 6 (23). The most common strain in China is the global subtype 1b (5), although subtype 6 is found at higher frequencies in southern China (27) and Hong Kong (42, 71).

HEPATITIS C AND HIV IN SOUTHEAST ASIA

HIV/HCV coinfection represents a public health problem in some regions of the Asia Pacific region. Studies from 10 countries in the Asia Pacific region comprising 89,452 HIV-positive individuals revealed discrepancies in prevalence rates in various countries. In Singapore, HCV coinfection prevalence is

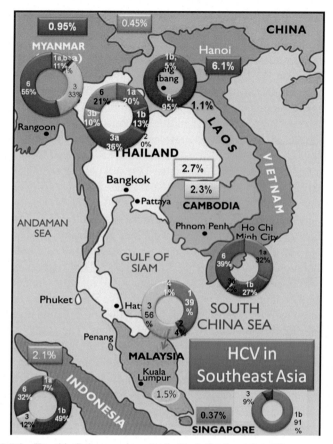

FIGURE 3.5.1   Hepatitis C virus genotype distribution in Southeast Asia [6].

3.8% [95% confidence interval (CI): 3.1–4.5]. In East Asia, HCV coinfection prevalence was 8.0% (95% CI: 6.4–9.8) in Hong Kong and 25.5% (95% CI: 17.5–34.4) in general HIV cohorts in China. In South Asia, HCV coinfection prevalence was 4.1% (95% CI: 1.7–7.3) in India and 42.6% (95% CI: 38.7–46.5) in Nepal. In Southeast Asia, HCV coinfection prevalence was 5.5% (95% CI: 4.9–6.1) in Cambodia, 5.3% (95% CI: 4.9–5.7) in Myanmar, and 5.1% (95% CI: 2.7–8.2) in Thailand, but higher in Vietnam (42.5%; 95% CI: 40.8–44.2) and Indonesia (17.9%; 95% CI: 15.0–20.9). The prevalence of HCV coinfection was higher in subpopulations of people who inject drugs [China 81.6% (95% CI: 74.1–88.0); Nepal 80.8% (95% CI: 76.4–84.9); Indonesia 81.6% (95% CI: 71.1–90.3)], former blood donors (China 82.9%; 95% CI: 73.9–90.3), and blood transfusion recipients (China 51.0%; 95% CI: 41.7–60.2). HCV coinfection prevalence within HIV populations is highly variable in the Asia Pacific region, between countries and in at-risk populations [9,10].

HEPATITIS C IN VIETNAM

In Vietnam, hepatitis C is more common than other Southeast Asian countries. In some studies, HCV prevalence rates range from 1% to 2.9% in the general population [11,12]. Recently, a large-scale HCV survey has been completed in Vietnam. A total of 8654 individuals with different risk factors were screened in five geographic regions: Ha Noi, Hai Phong, Da Nang, Khanh Hoa, and Can. The survey showed that military recruits and pregnant women had an HCV prevalence of 0.5%. This prevalence rate was significantly [12] lower than in many previous reports, in which the rates have ranged from 1% to 9%. Prevalence rates were significantly higher ($P<.001$) in intravenous drug users (55.6%), patients undergoing dialysis (26.6%), commercial sex workers (8.7%), and recipients of multiple blood transfusions (6.0%). The prevalence of HCV in patients undergoing dialysis varied but remained high in all regions (11% to 43%) and was associated with the receipt of blood transfusions [odds ratio (OR): 2.08 (1.85–2.34), $P=.001$], time from first transfusion [OR: 1.07 (1.01–1.13), $P=.023$], duration of dialysis [OR: 1.31 (1.19–1.43), $P<.001$], and male gender [OR: 1.60 (1.06–2.41), $P=.026$]. Another study [13] identifies hemodialysis as an increasingly common treatment in Vietnam. The study showed that 99% of patients reused their dialyzers and 46% had arteriovenous fistula on admission. HCV seropositivity was detected in 42% of patients. The study stressed the importance of safe practice for dialyzer reuse given the endemicity of hepatitis. We believe a national survey similar to ours about seroprevalence and infection control challenges would prepare Vietnam for providing safer satellite treatment units. Safe hemodialysis services should also comprise patient preparedness, education, and counseling [13].

Phylogenetic analysis revealed high genetic diversity, particularly among patients undergoing dialysis and multiple blood transfusions, identifying subtypes 1a (33%), 1b (27%), 2a (0.4%), 3a (0.7%), 3b (1.1%), 6a (18.8%), 6e (6.0%), 6h (4.6%), 6l (6.4%), and two clusters of novel genotype 6 variants (2.1%). HCV genotype 1 predominated in Vietnam (60%, n=169/282), but the proportion of infections attributable to genotype 1 varied between regions and risk groups, and, in the southern part of Vietnam, genotype 6 viruses dominated in patients undergoing dialysis and multiple blood transfusions (73.9%) [14].

HCV and HIV coinfection is frequent in Vietnam. The prevalence of HIV and HCV was 35.1% and 88.8%, respectively, and the prevalence of HIV/HCV coinfection and HCV monoinfection was 34.8% and 53.9%, respectively. After adjusting for confounders in multivariate analysis, ever reusing a syringe and needle was found to be significantly associated with HIV monoinfection [adjusted odds ratio (AOR): 3.13; 95% CI: 1.99–4.94] and HIV/HCV coinfection (AOR: 3.34; 95% CI: 2.02–5.51). Ever sharing diazepam or novocaine was also found to be significantly associated with HIV monoinfection (AOR: 2.14; 95% CI: 1.38–3.32) and HIV/HCV coinfection (AOR: 2.47; 95% CI: 1.57–3.90) [15].

HCV IN MYANMAR

In Myanmar, the prevalence of HCV infection is 2%, and HCV infection accounts for 25% of HCC in this country. HCV genotypes 1, 3, and 6 were observed in volunteer blood donors in and around the Myanmar city of Yangon [16]. Although there are several reports of HCV genotype 6 and its variants in Myanmar, the distribution of the HCV genotypes has not been well documented in areas other than Yangon. Previous studies showed that treatment with peginterferon and a weight-based dose of ribavirin for 24 or 48 week could lead to an 80%–100% sustained virologic response (SVR) rates in Myanmar. Current interferon-free treatments could lead to higher SVR rates (90%–95%) in patients infected with almost all HCV genotypes other than HCV genotype 3. In an era of heavy reliance on direct-acting antivirals against HCV, there is an increasing need to measure HCV genotypes, and this need will also increase specifically in Myanmar [17].

HCV IN INDONESIA

Some studies showed that the prevalence of anti-HCV in Java is higher than outside Java. In 1995 among 6971 blood donors in four big cities in Java, i.e., Jakarta, Bandung, Solo, and Surabaya, the prevalence of anti-HCV and HCV RNA was 1.5% and 1.1%, respectively, whereas from 8183 blood donors in 10 big cities outside Java, i.e., Medan, Palembang, and Padang (Sumatra island), Banjarmasin and Pontianak (Kalimantan), Manado and Makassar (Sulawesi island), Kupang and Dili (Timor island), and Ambon (Moluccas), the prevalence of anti-HCV and HCV RNA was 0.7% and 0.2%, respectively [18]. The high prevalence of HCV among blood donors is also reflected in the higher prevalence of anti-HCV in patients with chronic liver disease (CLD) in big cities in Java than big cities outside Java. In 1992, the prevalence of anti-HCV in patients with CLD (chronic hepatitis and liver cirrhosis) in Surabaya, Java island, was much higher than the prevalence in Mataram. Of the 343 patients with CLD in Dr. Soetomo General Hospital, Surabaya, and 114 patients with CLD in Mataram General Hospital, the prevalence of anti-HCV was much higher in Surabaya (61.8%) compared with those in Mataram (14.9%); conversely, HBV surface antigen (HBsAg) positives were only 27.7% in Surabaya and 41.2% in Mataram. Similarly, the prevalence of non-B–non-C hepatitis virus (NBNC) was much higher in patients with CLD in Mataram (43.9%) compared with those in Surabaya (9.6%). One of the possible reasons of high prevalence of NBNC in patients with CLD in Mataram was because of the existence of aflatoxin-contaminated food in the past [18,19].

The reported genotypes included in the HCV strains were classified into genotypes 1 (subtypes 1a, 1b, and 1c), 2 (subtypes 2a, 2e, and 2f), and 3 (subtypes 3a and 3k). The HCV 1b (47.3%) was the most prevalent, followed by subtypes 1c (18.7%), 3k (10.7%), 2a (10.0%), 1a (6.7%), 2e (5.3%), 2f (0.7%), and 3a (0.7%). HCV 1b was the most common in all patients, and the prevalence increased with the severity of liver disease [20].

Indonesia has a moderate-to-high rate of HBV infection and rapid epidemic growth of HIV infection; HCV infection can co-occur with HBV and HIV infections. In this study, 10 of 107 individuals (9.3%) were positive for HBsAg and/ or IIBV DNA, whereas 19 of 101 individuals (18.8%) with negative results for HBsAg were positive for HBV core antibody (anti-HBc). Seven of the 107 individuals (6.5%) were anti-HCV positive, and 16 of the 100 tested samples (16.0%) were HIV positive. Genotype and subtype analyses of all 10 HBV DNA (6 HBsAg positive and 4 anti-HBc positive) strains showed that 3 were of the HBV genotype/ HBsAg subtype C/adrq$^+$, 1 was of C/adw2, and 5 were of B/adw2. The HCV subtype distribution showed that 33.3% were of HCV-1b, and 66.7% were of HCV-3k (n=6). These distributions differed from those found in the general population of Surabaya, Indonesia. Interestingly, HIV subtype analysis showed a high prevalence of HIV, with possible recombinants of CRF01_AE and subtype B [21].

HCV IN PHILIPPINES

There are a few reports on the status of HCV in the Philippines. Few studies showed that antibodies against HCV were detected in 2.3% of commercial blood donors and in 23 (4.6%) of 502 sera from inmates in Metro Manila, the Philippines. The difference in the antibody prevalence between the two groups was statistically significant ($P<.05$). HCV RNA was detected in 78% of those with HCV-positive antibodies [22].

HCV IN CAMBODIA

Different surveys carried out in Cambodia found that the prevalence of viral hepatitis C ranged from 0.7% to 14.7%. It was higher (21%) in patients with HCC and in patients with high aspartate aminotransferase and alanine aminotransferase (39%) [23]. Another study showed that the prevalence of HCV RNA and anti-HCV was 2.3% and 5.8%, respectively. HCV RNA was detected in 39.3% of anti-HCV–positive samples and most of them were classified as genotype 6 (54.5%) and 1 (27.3%). History of operation and blood transfusion were significantly associated with the positivity for HBV infection and HCV RNA, respectively [24]. A case–control study investigated the routes of HCV transmission in people living with HIV/AIDS (PLHIV) in Cambodia. Cases were HCV/HIV co-infected patients (who tested reverse transcriptase–polymerase chain reaction positive for HCV RNA or had confirmed presence of HCV antibodies) (n=44). Controls were HIV monoinfected patients, with no HCV antibodies (n=160). They were recruited among the PLHIV presenting at one national reference center of HIV/AIDS. Multivariate analysis showed that factors associated with the co-infection were age older than 50 years (OR: 5.4, 95% CI: 1.5–19.6) and the exposure to multiple parenteral infusions before the year 2000 (OR: 3.4, 95% CI: 1.5–7.6), to surgery (OR: 2.6, 95% CI: 1.2–5.7), and to fibroscopy (OR: 2.4, 95% CI: 1.0–5.7). These results show the need to implement HCV screening in PLHIV, to support the implementation of national infection control guidelines, and to reinforce public awareness on the risks linked to parenteral medications [25].

REFERENCES

[1] Shepard CW, Finelli L, Alter MJ. Global epidemiology of hepatitis C virus infection. Lancet Infect Dis 2005;5:558–67.

[2] Sievert W, Altraif I, Razavi HA, Abdo A, Ahmed EA, et al. A systematic review of hepatitis C virus epidemiology in Asia, Australia and Egypt. Liver Int Off J Int Assoc Study Liver 2011;31(Suppl. 2):61–80.

[3] International Agency of Research Against Cancer. World cancer report. Geneva: IARC Publications; 2014.

[4] Puri P, Srivastava S. Lower chronic hepatitis B in South Asia despite all odds: bucking the trend of other infectious diseases. Trop Gastroenterol 2012;33:89–94.

[5] Nguyen H, Nguyen MH. Systematic review: Asian patients with chronic hepatitis C infection. Aliment Pharmacol Ther 2013;37:921–36.

[6] Wasitthankasem R, Vongpunsawad S, Siripon N, Suya C, Chulothok P, Chaiear K, Rujirojindakul P, Kanjana S, Theamboonlers A, Tangkijvanich P, Poovorawan Y. Genotypic distribution of hepatitis C virus in Thailand and Southeast Asia. PLoS One May 11, 2015;10(5):e0126764.

[7] Pybus O, Barnes E, Taggart R, Lemey P, Markov P, Rasachak B, Syhavong B, Phetsouvanah R, Sheridan I, Humphreys, Newton P, Klenerman P. Genetic history of hepatitis C virus In East Asia. J Virol 2009;83.

[8] Gower E, Estes C, Blach S, Razavi-Shearer K, Razavi H. Global epidemiology and genotype distribution of the hepatitis C virus infection. J Hepatol 2014;61:45–57.

[9] Martinello M, Amin J, Matthews GV, Dore GJ. Prevalence and disease burden of HCV coinfection in HIV cohorts in the Asia Pacific region: a systematic review and Meta-analysis. AIDS Rev April–June 2016;18(2):68–80.

[10] Durier N, Yunihastuti E, Ruxrungtham K, Kinh NV, Kamarulzaman A, Boettiger D, Widhani A, Avihingsanon A, Huy BV, Syed Omar SF, Sanityoso A, Chittmittrapap S, Dung NT, Pillai V, Suwan-Ampai T, Law M, Sohn AH, Matthews G. Chronic hepatitis C infection and liver disease in HIV-coinfected patients in Asia. J Viral Hepat March 2017;24(3):187–96.

[11] Dinford L, Carr M, Dean M, Waters A, Nguyen Ta Thi TH, Bui Thi LA, Duong Do H, Duong Thi TT, et al. Hepatitis C virus in Vietnam: high prevalence of infection in dialysis and multi-transfused patients involving diverse and novel virus variant. PLoS One 2012;7(8):e41266.

[12] Nakata S, Song P, Duc DD, Nguyen XQ, Murata K, et al. Hepatitis C and B virus infections in populations at low or high risk in Ho Chi Minh and Hanoi, Vietnam. J Gastroenterol Hepatol 1994;9:416–41.

[13] Duong CM, Olszyna DP, Nguyen PD, McLaws ML. Challenges of hemodialysis in Vietnam: experience from the first standardized district dialysis unit in Ho Chi Minh City. BMC Nephrol August 1, 2015;16:122.

[14] Pham DA, Leuangwutiwong P, Jittmittraphap A, Luplertlop N, Bach HK, Akkarathamrongsin S, Theamboonlers A, Poovorawan Y. High prevalence of Hepatitis C virus genotype 6 in Vietnam. Asian Pac J Allergy Immunol June–September 2009;27(2–3):153–60.

[15] Zhang L, Celentano DD, Le Minh N, Latkin CA, Mehta SH, Frangakis C, Ha TV, Mo TT, Sripaipan T, Davis WW, Quan VM, Go VF. Prevalence and correlates of HCV monoinfection and HIV and HCV coinfection among persons who inject drugs in Vietnam. Eur J Gastroenterol Hepatol May 2015;27(5):550–6.

[16] Myo-Khin, San-San-Oo Oo KM, Shimono K, Koide N, Okada S. Prevalence and factors associated with hepatitis C virus infection among Myanmar blood donors. Acta Med Okayama October 2010;64(5):317–21.

[17] Hlaing NK, Banerjee D, Mitrani R, Arker SH, Win KS, Tun NL, Thant Z, Win KM, Reddy KR. Hepatitis C virus therapy with peg-interferon and ribavirin in Myanmar: a resource-constrained country. World J Gastroenterol November 21, 2016;22(43):9613–22.

[18] Mulyanto. Epidemiology of hepatitis C in Indonesia. Paper presented in the 3rd International Eijkman Symposium, Jogjakarta, Indonesia September 30–October 3, 2004.

[19] Hotta H, Handayani R, Lusida MI, Soemarto W, Doi H, Miyajima H, Homma M. Subtypes analysis of hepatitis C virus in Indonesia on the basis of NS5b region sequences. J Clin Microbiol December 1994;32(12):3049–51.

[20] Utama A, Tania NP, Dhenni R, Gani RA, Hasan I, Sanityoso A, Lelosutan SA, Martamala R, Lesmana LA, Sulaiman A, Tai S. Genotype diversity of hepatitis C virus (HCV) in HCV-associated liver disease patients in Indonesia. Liver Int September 2010;30(8):1152–60.

[21] Hadikusumo AA, Utsumi T, Amin M, Khairunisa SQ, Istimagfirah A, Wahyuni RM, Lusida MI, Soetjipto, Rianto E, Juniastuti, Hayashi Y. High rates of hepatitis B virus (HBV), hepatitis C virus (HCV), and human immunodeficiency virus infections and uncommon HBV genotype/subtype and HCV subtype distributions among Transgender individuals in Surabaya, Indonesia. Jpn J Infect Dis November 22, 2016;69(6):493–9.

[22] Quesada P, Whitby D, Benavente Y, Miley W, Labo N, Chichareon S, Trong N, Shin HR, Anh PT, Thomas J, Matos E, Herrero R, Muñoz N, Molano M, Franceschi S, de Sanjosé S. Hepatitis C virus seroprevalence in the general female population from 8 countries. J Clin Virol July 2015;68:89–93.

[23] Buchy P, Monchy D, An TT, Srey CT, Tri DV, Son S, Glaziou P, Chien BT. Prevalence of hepatitis A, B, C and E virus markers among patients with elevated levels of alanine aminotransferase and aspartate aminotransferase in Phnom Penh (Cambodia) and Nha Trang (Central Vietnam). Bull Soc Pathol Exot August 2004;97(3):165–71.

[24] Yamada H, Fujimoto M, Svay S, Lim O, Hok S, Goto N, Ohisa M, Akita T, Matsuo J, Do SH, Katayama K, Miyakawa Y, Tanaka J. Seroprevalence, genotypic distribution and potential risk factors of hepatitis B and C virus infections among adults in Siem Reap, Cambodia. Hepatol Res April 2015;45(4):480–7.

[25] Goyet S, Lerolle N, Fournier-Nicolle I, Ken S, Nouhin J, Sowath L, Barennes H, Hak C, Ung C, Viretto G, Delfraissy JF, Khuon P, Segeral O. Risk factors for hepatitis C transmission in HIV patients, Hepacam study, ANRS 12267 Cambodia. AIDS Behav March 2014;18(3):495–504.

Hepatitis C Coinfections and Comorbidities in Developing Countries

Chapter 4.1

Hepatitis C and Schistosomiasis Coinfection

Sanaa M. Kamal
Ain Shams Faculty of Medicine, Cairo, Egypt

Chapter Outline

INTRODUCTION

Schistosomiasis is an acute and chronic parasitic disease caused by the trematode worms of the genus *Schistosoma* [1] (Fig. 4.1.1), which include intestinal and urogenital species. Schistosomiasis transmission has been reported from 78 countries and is endemic in 52 countries with moderate-to-high transmission [2,3] (Table 4.1.1, Fig. 4.1.2). *Schistosoma japonicum* is distributed in China, Indonesia, and the Philippines, whereas *Schistosoma mansoni* has a wider spread involving Africa, the Middle East, South America, and the West Indies [4,5]. *Schistosoma haematobium* has a distribution similar to that of *S. mansoni* but does not occur in South America or in the West Indies. In addition, *Schistosoma mekongi* and *Schistosoma intercalatum* are two species with local importance, causing intestinal schistosomiasis in the Mekong River basin of Southeast Asia and in Middle and West Africa, respectively [1–3,6].

Schistosomiasis caused by *S. mansoni* and *S. hematobium* is endemic in Egypt since ancient times. In 1972, the Egyptian Ministry of Health issued

Hepatitis C in Developing Countries. https://doi.org/10.1016/B978-0-12-803233-6.00010-2
107

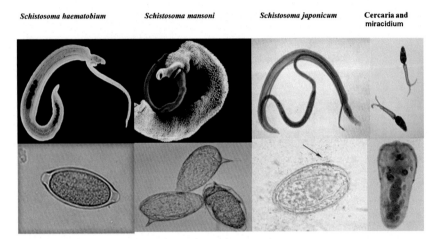

Schistosoma haematobium   *Schistosoma mansoni*   *Schistosoma japonicum*   Cercaria and miracidium

FIGURE 4.1.1   Schistosoma species.

TABLE 4.1.1 Prevalence of *Schistosoma haematobium* in Three Districts in Middle and Upper Egypt

	Beni Suef (%)	Menya (%)	Assiut (%)	Weighted Positive (%)	Estimated Prevalence
1977 Baseline	27.7	33.6	19.3	29.3	
1979	16.4	17.4	11.8	16.1	
1980	14.4	17.3	9.9	15.3	
1981	15.5	14.7	10.4	14.1	
1982	15.2	14.0	7.0	13.2	
1983	9.3	11.6	8.9	10.5	
1984	6.8	9.1	10.4	9.2	
1985	5.1	7.3	7.3	6.8	
1986	4.9	6.2	9.1	6.0	
1987	4.8	4.9	5.0	4.9	
1988	4.2	4.6	9.0	4.6	16.8
1989	2.6	4.6	2.7	3.9	14.8
1990	1.8	3.7	4.3	3.1	12.0
1991	1.7	3.4	3.2	2.9	10.7
1992	1.7	2.9	2.9	2.7	10.5
1993					11.2
1994					9.9

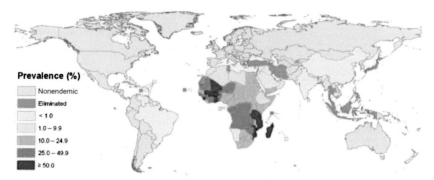

Prevalence (%)

Nonendemic
Eliminated
< 1.0
1.0 – 9.9
10.0 – 24.9
25.0 – 49.9
≥ 50.0

FIGURE 4.1.2 Worldwide prevalence of schistosomiasis.

Egypt's first official report on the burden of schistosomiasis. Since 1969, the government of Egypt has expanded its schistosomiasis control efforts to Upper and Middle Egypt and the Suez Canal area, relying heavily on foreign assistance. In 1984, the government spent just over 8% of the per capita public health expenditure on schistosomiasis control; in 1988, expenditure was cut to 5.2% [7].

The many years of efforts in Egypt have had an impact on schistosomiasis prevalence. According to a World Bank report, prevalence of schistosomiasis was reduced in Middle Egypt from about 30% in the late 1970s to about 10% in the late 1980s [7]. In Upper Egypt, the figures dropped from 21.7% positive samples in 1980 (in 775,000 persons examined) to 14.4% positive samples in 1988 (in over 3 million persons examined). The Ministry of Health reported that from 1982 to 1992 the prevalence of *S. haematobium* declined from about 15% to 1% in the Nile Delta and from 13% to 3% in Upper Egypt, and the prevalence of *S. mansoni* declined from about 40% to 20% in the Nile Delta [7–10].

SCHISTOSOMA LIFE CYCLE AND MODES OF TRANSMISSION

Transmission occurs when schistosomiasis-infected patients contaminate freshwater sources with their excreta containing parasite eggs, which hatch in water. People become infected when larval forms of the parasite – released by freshwater snails – penetrate the skin during contact with infested water (Fig. 4.1.3). In the body, the larvae develop into adult schistosomes. Adult worms live in the blood vessels where the females release eggs. Some of the ova are passed out of the body in the feces or urine to continue the parasite's lifecycle. Others become trapped in body tissues, causing immune reactions and progressive damage to organs [6]. Schistosomiasis results from immunologic reactions to *Schistosoma* ova trapped in tissues. Antigens released from the egg stimulate a granulomatous reaction involving T cells, macrophages, and eosinophils that results in

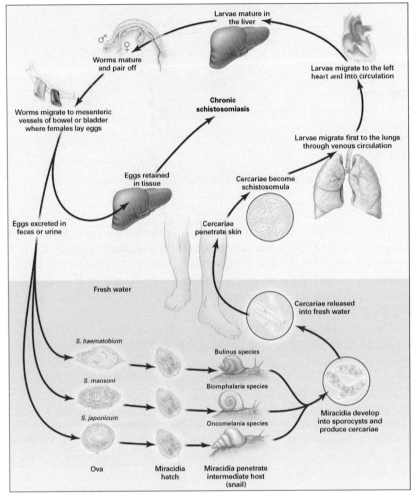

FIGURE 4.1.3 Life cycle of Schistosoma. *(Adapted from Allen et al. (2002).)*

clinical disease (Fig. 4.1.4). Symptoms and signs depend on the number and location of eggs trapped in the tissues. Initially, the inflammatory reaction is readily reversible. In the latter stages of the disease, the pathology is associated with collagen deposition and fibrosis, resulting in organ damage that may be only partially reversible [11] (Table 4.1.2).

ACUTE SCHISTOSOMIASIS

Acute schistosomiasis occurs most often in visitors to schistosome-endemic regions who are exposed to schistosome antigens for the first time at an older age than usual. It occurs weeks to months after infection, as a consequence of

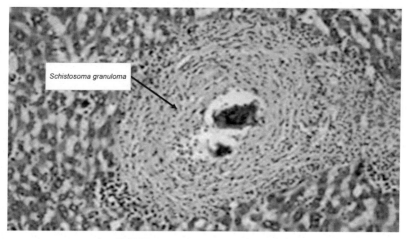

FIGURE 4.1.4 *Schistosoma granuloma.*

TABLE 4.1.2 Schistosoma Species and Geographical Distribution of Schistosomiasis

	Species	Geographical Distribution
Intestinal schistosomiasis	*Schistosoma mansoni*	Africa, the Middle East, the Caribbean, Brazil, Venezuela, and Suriname
	Schistosoma japonicum	China, Indonesia, the Philippines
	Schistosoma mekongi	Several districts of Cambodia and the Lao People's Democratic Republic
	Schistosoma guineensis and related *Schistosoma intercalatum*	Rain forest areas of central Africa
Urogenital schistosomiasis	*Schistosoma haematobium*	Africa, the Middle East, Corsica

worm maturation, egg production, release of egg antigen, and the host's florid granulomatous and immune complex responses [6]. Acute schistosomiasis is sometimes referred to as Katayama syndrome and the typical clinical presentation is a sudden onset of fever, malaise, myalgia, headache, eosinophilia, fatigue, and abdominal pain lasting 2–10 weeks [6,11,12]. The limited presentation of this syndrome in residents of endemic regions is probably a result of *in utero* priming of T-lymphocyte and B-lymphocyte responses of babies born to mothers with helminthic infections [6,12].

CHRONIC SCHISTOSOMIASIS

The manifestations of the chronic phase of the infection are mostly because of the granulomatous inflammatory reaction against the schistosome ova deposited in different organs and tissues [7]. In intestinal schistosomiasis, egg deposition occurs mainly in the liver and the intestinal wall and can lead to multiple-granuloma formation and tissue lesions in these organs [11]. This causes intestinal mucosal hyperplasia, polyposis, and ulceration, which is clinically reflected as abdominal pain, chronic diarrhea, and rectal bleeding [13,14]. Granulomas in the liver results in a periportal fibrosis extending to advanced disease, with portal hypertension and hepatosplenomegaly. Ascites and variceal bleeding are two serious and common complications at this stage, which can result in the death of the patient [15].

In urinary schistosomiasis, granulomas form after deposition of ova mainly in the urinary bladder wall, resulting in abnormalities in the mucosa [16]. Patients infected with *S. hematobium* develop lower urinary tract symptoms, such as hematuria, frequency, and dysuria. Further complications include bladder calcification, urinary tract fibrosis causing obstructive uropathy, and bladder malignancies [16–18]. *S. haematobium* is a class 1 carcinogen and has also been shown to increase the risk of sexually transmitted infections, including HIV infection, especially in female genital schistosomiasis haematobia [16,18].

Altogether, the spectrum of clinical manifestations and complications in schistosomiasis mainly depend on the intensity of the infection and the magnitude of the host immune response. Once the disease is at an advanced stage, significant morbidity with lifelong disabilities or severe complications resulting in death can occur.

IMMUNE RESPONSES IN SCHISTOSOMIASIS

Morbidity in humans infected with *S. mansoni* results primarily from deposition of parasite ova in the portal areas, inducing a T-cell–dependent granulomatous response that progresses to irreversible fibrosis and severe portal hypertension in more than 60% of cases [16]. The murine models of this disease have been widely studied because they permit a variety of genetic and other experimental approaches. The murine infections differ from the human, monkey, and baboon infections in many ways, especially in the number of adult worms per unit of body weight and the distribution of ova between the liver and mesenteric circulation. However, all these species develop hepatic granulomatous (HG)

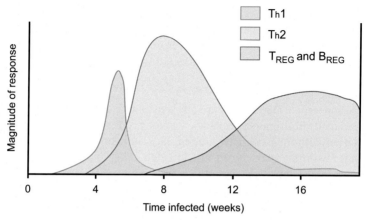

FIGURE 4.1.5   Immune responses in schistosomiasis.

inflammations that have similar dynamics and cellular compositions and are spontaneously downmodulated. Diverse evidence strongly implicates both Th1 [interleukin-2 (IL-2) and interferon-γ (IFN-γ)] and Th2 (IL-4, IL-13, IL-5, and tumor necrosis factor-α) cytokines in the regulation of HG size [19,20]. The high concentration of soluble egg molecules exerts multiple levels of control over the development of host immune responses, with predominant early IFN-γ production followed by dominant Th2 responses 8 weeks thereafter. Soluble egg antigen was also found to inhibit the ability of dendritic cells to make IL-12 and induces Th2-polarized adaptive immune responses that in combination with regulatory T-cell responses effectively limit Th1 response development [21] (Fig. 4.1.5).

IMMUNE RESPONSES IN HEPATITIS C INFECTION

In terms of the immune response in hepatitis C virus (HCV) infection, the infection usually persists in part because of the rapid replication of the virus and its tendency to mutate and form variants known as quasi-species that cannot be contained by the immune response. Although antibodies may neutralize some viral quasi-species, anti-HCV antibodies do not protect from subsequent HCV infections [22,23]. HCV infection induces CD4+ and CD8+ T–cell (CTL) responses. Several studies in humans and chimpanzees have demonstrated a significant association between a strong and sustained virus-specific CD4+ and CD8+ T-cell responses and spontaneous viral clearance in acute HCV [24,25]. Individuals who fail to mount or sustain such responses usually develop persistent viremia and chronic infection [26,27]. Two important observations are key to understanding the failure of the immune response in the majority of infected individuals. First, sudden failure of the HCV-specific CD4+ helper T-cell response is associated with a rebound in viremia and a chronic disease course. Second, HCV-specific CD4+ T cells are rarely detected in persistently infected individuals, and although CD8+ T cells are more readily detectable, they are somewhat impaired in many effector functions [22,28].

Virus-specific CD8+ T cells are arrested at an early differentiation stage, have reduced proliferative capacity, cannot produce IFN-γ in response to stimulation with their cognate antigen, lack stores of intracellular perforin associated with the capacity to kill infected cells, and produce regulatory cytokines that dampen the immune response [22,29]. Whether such functional impairments occur early and cause the persistence or are merely a consequence of persistent viremia is unknown. The essential role of CD4+ and CD8+ T cells in HCV clearance is underscored by results of antibody-mediated depletion studies in the chimpanzee. Most important, depletion of CD4+ T cells results in an inadequate CD8+ T-cell response and facilitates mutations in targeted CTL epitopes and virus persistence, suggesting a central role for CD4+ helper T cells in sustaining an efficient antiviral CTL response [30–33] (Fig. 4.1.6).

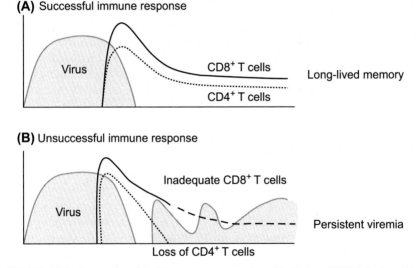

FIGURE 4.1.6 Proposed model of immune response to hepatitis C virus (HCV) infection. (A) Successful immune response to HCV is characterized by a robust, persistent, virus-specific CD4+ and CD8+ T-cell–mediated immune response that leads to control of viremia. Maintenance of virus-specific CD4+ and CD8+ T cells beyond this point results in permanent clearance of the infection. (B) An unsuccessful immune response is transient in nature. Virus-specific CD4+ and CD8+ T cells are generated and lead to partial control of viral replication, but the responses wane and become ineffective.

CONCOMITANT SCHISTOSOMIASIS AND HCV INFECTION

Concomitant schistosomiasis and HCV infection is common in Egypt and other developing countries [34]. Patients with concomitant HCV and schistosomiasis exhibit a unique clinical, virological, and histological pattern manifested by virus persistence with high HCV RNA titres, higher necroinflammatory and fibrosis scores in their liver biopsies, and markedly accelerated disease course once chronic HCV infection is established [34].

EFFECT OF SCHISTOSOMIASIS ON THE OUTCOME OF ACUTE HEPATITIS C INFECTION

Immune responses during the first few months of acute HCV infection seem crucial for viral control. To determine whether immunologic mechanisms are responsible for this alteration in the natural history of HCV, Kamal et al. prospectively investigated the HCV-specific CD4+ and cytokine responses in patients with acute HCV hepatitis with or without *S. mansoni* coinfection and correlated these responses to the outcome of acute HCV infection [35]. Whereas 30% of patients infected with HCV alone recovered from acute HCV, all patients with *S. mansoni* coinfection progressed to histologically proven chronic hepatitis. Coinfected patients had either absent or transient weak HCV-specific CD4+ responses with Th0/Th2 cytokine production. These findings show that patients with acute HCV and schistosomiasis coinfection cannot clear viremia and cannot generate an efficient, broad, HCV-specific CD4+/Th1 T-cell response, which might play a role in the persistence and severity of HCV infection in patients with coinfection [35].

EFFECT OF SCHISTOSOMIASIS ON THE PROGRESSION OF CHRONIC HEPATITIS C

Patients with established chronic HCV and *S. mansoni* coinfection fail to mount significant HCV-specific CD4+ T-cell responses and show alteration in the cytokine milieu along with more severe liver disease [35,36]. It has been also shown that T cells from persons with schistosomiasis have reduced IFN-γ–, IL-4–, and IL-10–secreting HCV-specific T-cell responses [35,36]. Elrefaei et al. characterized the HCV-specific CD8+ T-cell responses in a cohort of HCV-infected individuals with or without *S. mansoni* coinfection. They observed a significantly decreased CD27− CD28− (late differentiated) memory T-cell population in the HCV coinfected individuals compared with those with HCV infection alone. In contrast, there was no significant difference in the CD27+ CD28+ (early differentiated) memory T cells between the two groups [37]. Given the more rapid progression of liver disease in these subjects, HCV/*S. mansoni* coinfection represents a useful model for examining the relationship between the immune response and subsequent liver injury. In a longitudinal prospective study, Kamal et al. correlated the kinetics of peripheral and intrahepatic CD4+ responses to the rate of hepatic fibrosis progression in patients with HCV/*S. mansoni* coinfection and patients with HCV monoinfection [36]. Coinfected patients had accelerated liver fibrosis, with rates of progression of fibrosis of >0.58 fibrosis units/year; this rate compared unfavorably with the 0.1 fibrosis units/year seen in the subjects with HCV infection alone, despite the comparable necroinflammatory and fibrosis scores in both groups at the early phase of chronic HCV infection. The two groups of patients differed sharply in their CD4+ T-cell responsiveness to HCV. The frequency, magnitude, and breadth of the HCV-specific T-cell responses in subjects with HCV infection alone were significantly higher than those in coinfected subjects. The study identified a significant inverse association between intrahepatic CD4+ T-cell responses and progression of fibrosis, suggesting that absent or diminished HCV-specific CD4+ T-cell responses in the liver, particularly during the early phase of chronic evolution, favor liver damage and progression of disease [38,39] (Figs. 4.1.7 and 4.1.8).

In summary, patients with HCV and schistosomiasis have unique features characterized by very low rates of spontaneous resolution of acute hepatitis C, high levels of viremia, rapid rate of progression of fibrosis, and cirrhosis.

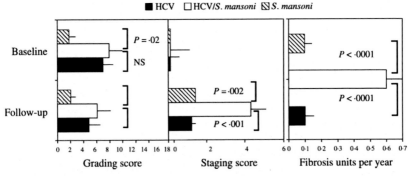

FIGURE 4.1.7 Liver histological assessment and progression of fibrosis in monoinfected and coinfected subjects. Left panel, comparison of the necroinflammatory scores at baseline biopsies (biopsy 1, performed 6–10 months after acute hepatitis) and follow-up biopsies (biopsy 2, performed 8–10 years later), in 23 monoinfected subjects (*filled bars*), 25 coinfected subjects (*open bars*), and 20 control subjects with *S. mansoni* infection alone (*striped bars*). *Bars* represent means, and *lines* represent SD. Middle panel, fibrosis scores at baseline biopsies and follow-up biopsies. At baseline liver biopsies, monoinfected subjects, coinfected subjects, and control subjects had no fibrosis (stage 0). Coinfected subjects had significantly greater increases in fibrosis scores detected in follow-up biopsies, compared with monoinfected subjects. Right panel, rates of progression of fibrosis. The rate of liver progression of fibrosis was significantly higher in coinfected subjects than in monoinfected subjects (0.58±0.13 in the coinfected group vs. 0.1±0.06 in the monoinfected group; *P*=.001) [39].

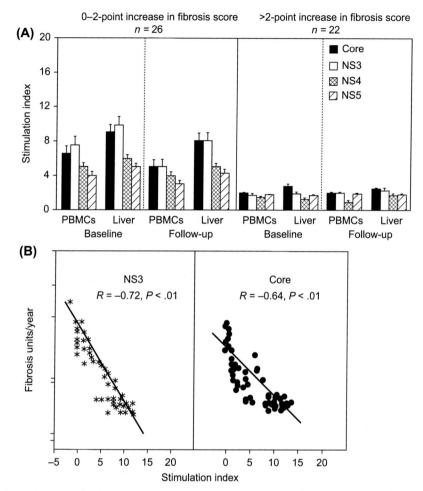

FIGURE 4.1.8 Peripheral and intrahepatic hepatitis C virus (HCV)–specific CD4+-proliferative T-cell responses in relation to progression of fibrosis. (A) CD4+ T-cell responses to four HCV proteins (HCV core, NS3, NS4, and NS5), expressed as stimulation index (Y-axis), in subjects with a 0- to 2-point increase in fibrosis score (left) and subjects with a 12-point increase (right) at the follow-up biopsies. (B) Relation of baseline intrahepatic HCV (NS3) and HCV core-specific CD4+-proliferative T-cell responses to progression of fibrosis (fibrosis units per year): the two parameters show a significant inverse correlation ($R = -0.72$; $P > .01$; $R = -0.64$; $P < .01$, respectively).

REFERENCES

[1] Schistosomiasis, Fact Sheet No 115. World Health Organization; February 2010. Available from: http://www.who.int/mediacentre/factsheets/fs115/en/.

[2] Chistulo L, Loverde P, Engels D. Disease Watch: schistosomiasis. TDR Nat Rev Microbiol 2004;2:12.

[3] Mari L, Gatto M, Ciddio M, Dia ED, Sokolow SH, De Leo GA, Casagrandi R. Big-data-driven modeling unveils country-wide drivers of endemic schistosomiasis. Sci Rep March 28, 2017;7(1):489.

[4] Gryseels B, Polman K, Clerinx J, Kestens L. Human schistosomiasis. Lancet 2006;368:1106–18. http://dx.doi.org/10.1016/S0140-6736(06)69440-3.

[5] Barakat RMR. Epidemiology of schistosomiasis in Egypt: travel through time. J Adv Res 2013;4:425–32. http://dx.doi.org/10.1016/j.jare.2012.07.003.

[6] Jordan P, Webbe G, Sturrock R. Human schistosomiasis. Wallingford, England: CAB; 1993.

[7] Cochrane D, Bernhard G, Liese H. Egypt: schistosomiasis control. In: Liese Bernhard H, Sachdeva Paramjit S, Glynn Cochrane D, editors. Organizing and managing tropical disease control programmes: case studies. Washington, DC.: The World Bank; 1992. p. 37–50.

[8] Khoby El, Taha, Galal N, Fenwick A. Schistosomiasis control in Egypt. Schistosomiasis Res Proj 1993;3(3):2.

[9] El Malatawy A. An integrated approach to schistosomiasis control. [Unpublished document]. Cairo: UNICEF; January 1989.

[10] El Malatawy A, El Habashy A, Lechine N, Dixon H, Davis A, Mott KE. Selective population chemotherapy among schoolchildren in Beheira Governorate: the UNICEF/Arab Republic of Egypt/WHO schistosomiasis control Project. Bull World Health Organ 1992;70:47–56.

[11] Warren K. The pathology of schistosome infections. Helminthol Abstr Ser [A] 1973;42:590–633.

[12] Doherty JF, Moody AH, Wright SG. Katayama fever: an acute manifestation of schistosomiasis. BMJ 1996;313:1071–2.

[13] Gryseels B, Polman K, Clerinx J, Kestens L. Human schistosomiasis. Lancet September 23, 2006;368(9541):1106–18.

[14] Cheever AW. A quantitative post-mortem study of Schistosomiasis mansoni in man. Am J Trop Med Hyg 1968;17:38–64.

[15] Strauss E. Hepatosplenic schistosomiasis: a model for the study of portal hypertension. Ann Hepatol 2002;1:6–11.

[16] Burke ML, Jones MK, Gobert GN, Li YS, Ellis MK, McManus DP. Immunopathogenesis of human schistosomiasis. Parasite Immunol 2009;31:163–76.

[17] King CH, Keating CE, Muruka JF, et al. Urinary tract morbidity in schistosomiasis haematobia: associations with age and intensity of infection in an endemic area of Coast Province, Kenya. Am J Trop Med Hyg 1988;39:361–8.

[18] Barsoum RS. Urinary schistosomiasis: review. J Adv Res September 2013;4(5):453–9.

[19] Pearce EJ, MacDonald AS. The immunobiology of schistosomiasis. Nat Rev Immunol 2002;2:499–511.

[20] Fairfax K, Nascimento M, Huang SC, Everts B, Pearce EJ. Th2 responses in schistosomiasis. Semin Immunopathol 2012;34:863–71.

[21] Peterson WP, Von Lichtenberg F. Studies on granuloma formation. IV. In vivo antigenicity of schistosome egg antigen in lung tissue. J Immunol 1965;95:959–65.

[22] Shoukry NH, Cawthon AG, Walker CM. Cell-mediated immunity and the outcome of hepatitis C virus infection. Annu Rev Microbiol 2004;58:391–424.

[23] Hartling HJ, Ballegaard VC, Nielsen NS, Gaardbo JC, Nielsen SD. Immune regulation in chronic hepatitis C virus infection. Scand J Gastroenterol July 19, 2016:1–11.

[24] Ishii S, Koziel MJ. Immune responses during acute and chronic infection with hepatitis C virus. Clin Immunol August 2008;128(2):133–47.

[25] Burke Schinkel SC, Carrasco-Medina L, Cooper CL, Crawley AM. Generalized liver- and blood-Derived CD8+ T-cell impairment in response to cytokines in chronic hepatitis C virus infection. PLoS One June 17, 2016;11(6):55.

[26] Chen Yi Mei SL, Burchell J, Skinner N, Millen R, Matthews G, Hellard M, Dore GJ, Desmond PV, Sundararajan V, Thompson AJ, Visvanathan K, Sasadeusz J. Toll-like receptor expression and signaling in peripheral blood mononuclear cells correlate with clinical outcomes in acute hepatitis C virus infection. J Infect Dis June 9, 2016.

[27] Kamal SM. Acute hepatitis C: a systematic review. Am J Gastroenterol 2008;103(5):1283–97.

[28] Goh CC, Roggerson KM, Lee HC, Golden-Mason L, Rosen HR, Hahn YS. Hepatitis C virus-induced myeloid-derived suppressor cells suppress NK cell IFN-γ production by altering cellular metabolism via Arginase-1. J Immunol March 1, 2016;196(5):2283–92.

[29] Serti E, Park H, Keane M, O'Keefe AC, Rivera E, Liang TJ, Ghany M, Rehermann B. Rapid decrease in hepatitis C viremia by direct acting antivirals improves the natural killer cell response to IFNα. Gut 2016;2015–(31):33.

[30] Quarleri JF, Oubiña JR. Hepatitis C virus strategies to evade the specific-T cell response: a possible mission favoring its persistence. Ann Hepatol 2016 Jan-Feb;15(1):17–26.

[31] Raziorrouh B, Sacher K, Tawar RG, Emmerich F, Neumann-Haefelin C, Baumert TF, Thimme R, Boettler T. Virus-specific CD4+ T cells have functional and phenotypic characteristics of follicular T-helper cells in patients with acute and chronic HCV. Infect Gastroenterol March 2016;150(3):696–770.

[32] Kokordelis P, Krämer B, Boesecke C, Voigt E, Ingiliz P, Glässner A, Wolter F, Srassburg CP, Spengler U, Rockstroh JK, Nattermann J. CD3(+) CD56(+) natural killer-like T cells display anti-HCV activity but are functionally impaired in HIV(+) patients with acute hepatitis C. J Acquir Immune Defic Syndr December 1, 2015;70(4):338–46.

[33] Chen AY, Hoare M, Shankar AN, Allison M, Alexander GJ, Michalak TI. Persistence of hepatitis C virus traces after spontaneous resolution of hepatitis C. PLoS One October 16, 2015;10(10):e0140312.

[34] Kamal S, Madwar M, Bianchi L, Tawil AE, Fawzy R, Peters T, Rasenack JW. Clinical, virological and histopathological features: long-term follow-up in patients with chronic hepatitis C co-infected with S. mansoni. Liver July 2000;20(4):281–9.

[35] Kamal SM, Bianchi L, Al Tawil A, Koziel M, El Sayed Khalifa K, Peter T, Rasenack JW. Specific cellular immune response and cytokine patterns in patients coinfected with hepatitis C virus and Schistosoma mansoni. J Infect Dis October 15, 2001;184(8):972–82.

[36] Kamal SM, Rasenack JW, Bianchi L, Al Tawil A, El Sayed Khalifa K, Peter T, Mansour H, Ezzat W, Koziel M. Acute hepatitis C without and with schistosomiasis: correlation with hepatitis C-specific CD4(+) T-cell and cytokine response. Gastroenterology September 2001;121(3):646–56.

[37] Elrefaei M, El-Sheikh N, Kamal K, Cao H. HCV-specific CD27- CD28- memory T cells are depleted in hepatitis C virus and Schistosoma mansoni co-infection. Immunology 2003;110:513–8.

[38] Kamal SM, Graham CS, He Q, Bianchi L, Tawil AA, Rasenack JW, Khalifa KA, Massoud MM, Koziel MJ. Kinetics of intrahepatic hepatitis C virus (HCV)-specific CD4+ T cell responses in HCV and Schistosoma mansoni coinfection: relation to progression of liver fibrosis. J Infect Dis April 1, 2004;189(7):1140–50.

[39] Kamal SM, Turner B, He Q, Rasenack J, Bianchi L, Al Tawil A, Nooman A, Massoud M, Koziel MJ, Afdhal NH. Progression of fibrosis in hepatitis C with and without schistosomiasis: correlation with serum markers of fibrosis. Hepatology April 2006;43(4):771–9.

Chapter 4.2

Hepatitis C and Helminthic Infections

Khalifa S. Khalifa[1], Othman Amin[2]
[1]Ain Shams Faculty of Medicine, Cairo, Egypt; [2]Al Khartoum University, Khartoum, Sudan

INTRODUCTION

Hepatitis C and helminthes are infectious diseases that affect millions of people worldwide. The prevailing immune responses evoked by helminthic and protozoal infections influence the outcome of hepatitis C infection. Plasmodium parasite for malaria and hepatitis C virus (HCV) for hepatitis C share some similarities in their development within the hepatocytes of the liver. In several countries where both infections are prevalent, coinfection of these two pathogens is common. Plasmodium parasites and HCV utilize four common host entry factors to gain entry into hepatocytes: heparan sulfate proteoglycans, scavenger receptor-B1, cluster of differentiation 81, and apolipoprotein E (apoE). ApoE incorporated into new HCV virions plays a key role in viral infectivity.

Hepatitis C in Developing Countries. https://doi.org/10.1016/B978-0-12-803233-6.00011-4

121

Understanding the immune interactions in HCV and helminthic coinfections provides important insights into the pathogenesis of the emerging pathologies.

CHARACTERISTICS OF IMMUNE RESPONSE TO HELMINTHS

Helminthic Infections Induce a Th2-Biased Immune Response in Their Hosts

Progression of infectious diseases, including parasitic diseases, often depends on which cytokine profile, type 1 or type 2, is preferentially induced [1]. In the majority of cases, the immune responses of the hosts to helminthic infection are remarkably similar, being Th2-like with the production of significant quantities of interleukin (IL)-4, IL-5, IL-9, IL-10, and IL-13 and consequently the development of strong immunoglobulin E (IgE), eosinophil, and mast cell responses. Allied to this, polarization of the immune response to Th2 also downregulates the Th1 cell subset [1–7]. Despite the vast amount of research investigating how the mammalian host recognizes helminth antigen (Ag) and reacts with a Th2 response, this issue remains unresolved. There are multiple suggested pathways for the initiation of the Th2 response in which IL-4 produced early in the response plays an important role in the consolidation and amplification of the Th2 pathway [8,9]. Candidate sources of IL-4 include atypical subsets of T cells not restricted by class II major histocompatibility complex, including NK1+ CD4+ T cells, conventional CD4+ naïve and memory T cells [8,10–12], eosinophils, cells of the mast cell/basophil lineage [13–15], antigen-presenting cells, and dendritic cells [6].

Chronic Immune Activation, Hyporesponsiveness, and Anergy Associated With Helminthic Infections

As part of their adaptation to a parasitic life cycle, helminths have developed mechanisms to overcome the host's immune response. Such mechanisms contribute to the chronicity of infections and their persistence in adult populations. There is accumulating evidence that such prolonged confrontation between helminths and the host results in a state of chronic immune activation associated with hyporesponsiveness and anergy [16,17]. Several observations support this concept; for example, the initial marked T-cell responsiveness following filarial and schistosomal infection is diminished in most patients as the infection becomes chronic [18–20]. People chronically infected with the filarial parasite *Onchocerca volvulus* have depressed cellular reactivity in vitro and deficient production of IL-2 in response to *O. volvulus*–specific antigenic stimulation [21]. Subjects infected with *Schistosoma mansoni* have impaired tetanus toxoid–specific immune responses [14]. Helminths are also known to impede the host's capacity to mount an effective immune response after vaccination [14,22–26].

The underlying mechanisms behind the hyporesponsiveness and anergy during helminthic infections seem multifactorial and might include increased threshold for effective immune activation of T cells; defective intracellular signaling; a decrease in the number of costimulatory molecules; an increase in the number of T regulatory (Treg) cells; an increase in the amount of intracellular-negative regulator of T-cell activation in T cells; dysregulation of cytokine secretion; and a T-cell imbalance (reviewed in detail in [17]). Host macrophages may alternatively be activated by the parasite to effect suppression via a contact-dependent mechanism [27,28] and via the production of interferon (IFN)-γ and NO (nitrous oxide) in response to parasite glycoconjugates [29]. Another mechanism of anergy induced by worms is selective upregulation of programmed death ligand 1, which interacts with programmed death 1 in T cells, resulting in anergy [30].

Immunoregulation During Helminthic Infections

Helminthic infections induce strong IgE responses, which in combination with high Ag levels would be expected to lead to allergic symptoms and possibly anaphylaxis [31]. However, helminth-infected individuals rarely have allergic reactions to these parasites and moreover appear to suffer less from allergic disorders in general than do helminth-free individuals [32,33]. Several investigators found that regular anthelminthic treatment of children with intestinal helminthic infection resulted in a significant decrease in the elevated total serum IgE levels and increase in both skin test reactivity and serum levels of specific IgE against environmental allergens, indicating that helminths directly suppress allergic reactions [34,35]. On the other hand, other investigators found that *Ascaris lumbricoides* infection was associated with increased risk of childhood asthma and atopy [36] and that regular anthelminthic treatment of asthmatic patients with albendazole for 1 year resulted in a significant improvement in all the indicators of clinical status, not only for the period of anthelminthic administration but also for the year following [37]. The negative association between helminthic infections and allergic diseases may be because of the production of high levels of polyclonal IgE that cause saturation of mast cell Fc epsilon receptors [38,39] or by promoting high levels of regulatory cytokines capable of downregulating allergic responses [5]. The suppression of atopy by helminth infections seems to depend on a high burden of parasites, while low-level infections, as a result of treatment or low transmission, can stimulate rather than suppress allergic reactivity [33,36], probably by enhancing Th2 responses without sufficiently stimulating the regulatory network [40,41].

Thus, it appears that helminths regulate the host immune system, and multiple mechanisms have been proposed for such regulation [2]. Cytokines such as IL-10 and transforming growth factor β (TGF-β), and Tregs and suppressor macrophages, are involved in this regulation [28]. Indeed, it is clear that IL-10 plays a major role in the regulation of the intensity of both Th1 and Th2 responses during helminthic infections [42,43]. For instance, IL-10 inhibits macrophage and dendritic cell function and suppresses the production of important proinflammatory cytokines such as tumor necrosis factor alpha, IL-12, IL-1, and nitric oxide and various chemokines, inhibiting development of both Th1 and Th2 cells. Several studies support the concept that it is not the Th1 to Th2 shift but rather other cytokines, primarily IL-10 and TGF-β, that mediate the antigen-specific hyporesponsiveness characteristic of chronic human or primate helminthic infections

[44–47]. Such hyporesponsiveness and anergy could be tied to the effects of Treg/suppressor cells present in these situations. These cells constitute 5%–10% of peripheral CD4$^+$ T cells in naive mice and humans and suppress several potentially pathogenic responses in vivo, particularly T-cell responses directed to self-antigens. There are three phenotypes of T cells with predominant regulatory function: the natural Treg, Th3, and Tr1 subtypes. By now it has become clear that Treg cells belong to a population of CD4$^+$ T cells that co-express CD25 (the IL-2 receptor alpha-chain), constitutively express CTLA-4 (cytotoxic T-lymphocyte-associated protein *4*), and often secrete IL-10, TGF-β, IFN-α, and IL-5 and the cellular transcription factor FOXP3 [48,49]. Th3 cells make TGF-β, which inhibits development of both Th1 and Th2 cells [46]. Tr1 cells make copious amounts of IL-10, which inhibits both Th1 and Th2 responses [50].

CONCURRENT HELMINTHIC AND HEPATITIS C INFECTIONS

The biased Th2 immune response and hyporesponsiveness associated with chronic helminthiases might not only impede the capacity of infected individuals to cope with concurrent infectious pathogens but also their capacity to mount an effective immune response after vaccination [14,22–24]. Several observations support this concept. Studies in laboratory animals showed that mice infected with *S. mansoni* for different durations mount diminished quantitative and qualitative serum anti-toxin levels to diphtheria toxoid when compared with noninfected vaccinated mice [24]. Also, mice infected with *S. mansoni* show depressed mycobacterial, antigen-specific Th1-type responses [51,52]. Similarly, *S. mansoni*–infected patients, particularly those with hepatointestinal disease, show a diminished ability to mount an immune response to *Salmonella typhi* after immunization with a typhoid vaccine [23]. Populations harboring preexisting trematode or nematode infections have diminished protective immunity to tuberculosis (TB) induced by vaccination [53,54]. Significant improvement in mycobacterial-specific immune responses occurs after anthelminthic therapy [25]. In humans, concurrent infections with helminths such as *S. mansoni* and *Onchocerca* diminished the magnitude of the Th1 immune response to tetanus toxoid [14,55]. *A. lumbricoides* was found to diminish the Th1 cytokine response (IL-2 and IFN-γ) to recombinant cholera toxin B subunit after vaccination with the live oral cholera vaccine, whereas healthy individuals and *A. lumbricoides*–infected subjects who were pretreated with albendazole before vaccination produced a similar Th1-predominated response [26]. These observations may explain the weak responsiveness to and lessened success of all or most vaccines in developing countries compared with developed countries [22,25,56].

Nevertheless, the impact of helminths on the outcome of vaccination is not yet established. As the failure of vaccination is mostly attributed to the stimulated Th2 and diminished Th1 responses, by the same logic the increase in type 2 cytokines occurring immediately after deworming [57–59] might impair the response to other unrelated antigens, including vaccines. Treatment of *S. mansoni* in HIV coinfected Ugandan adults resulted in increased helminth-specific type 2 cytokine responses and HIV-1 load and diminished circulating IL-10 concentrations [57].

However, Elliott et al. found that treatment of Ugandan pregnant women with a single dose of albendazole during the second trimester of pregnancy enhanced the Th1 responses of the treated mothers and suppressed responses of their 1-year-old infants to BCG stimulation, whereas there was a "better" response to BCG in the infants of untreated mothers with light-to-moderate hookworm infection [60].

The usefulness of deworming is beyond dispute and has gained support from international health organizations. It is believed that such an approach would have a big impact on a number of health parameters, such as general morbidity, anemia, growth, and learning abilities [61–64] and on improving the outcome of vaccination [56]. However, the strategy of mass deworming needs proper evaluation to determine the appropriate timing of anthelminthic treatment in relation of vaccination.

EFFECT OF HELMINTHS ON VIRAL INFECTIONS

Concomitant human helminthic and viral infections are frequent in many geographical areas, particularly in tropical and subtropical countries. Several examples exist, such as human coinfection with *S. mansoni* and hepatitis B and hepatitis C viruses, which is associated with increased hepatic viral burdens and severe liver pathology. Several helminthic infections have been reported concurrent with HIV infection, with some studies showing an impact on the natural history and progression of the HIV infection. Likewise, studies in animal models have demonstrated a negative impact of helminths on the outcome of virus infections. One group found that BALB/c mice infected 7 weeks previously with *S. mansoni* and challenged with a recombinant vaccinia virus vPE16 expressing the HIV envelope protein gp160 showed a marked delay in hepatic viral clearance compared with mice infected with vPE16 alone. This study demonstrated enhanced viral replication within the egg granulomas and this combined with virus-specific cytokine skewed toward Th2 seem to promote the expansion of vaccinia infection [65]. Similar events apply to other viral infections. In an *S. mansoni*/lymphocytic choriomeningitis virus (LCMV) coinfection model, viral coinfection during the Th2-dominated granulomatous phase of the schistosome infection resulted in induction of a strong LCMV-specific T-cell response, with infiltration of high numbers of LCMV-specific IFN-γ–producing CD8$^+$ cells into the liver and rapid increase in morbidity and hepatotoxicity. Interestingly, the liver of coinfected mice was extremely susceptible to viral replication with a reduced intrahepatic type I IFN response after virus infection. These results suggest that suppression of the antiviral type I IFN response by schistosome egg antigens in vivo predisposes the liver to enhanced viral replication with ensuing immunopathologic consequences [66].

Patients with HCV and schistosomiasis are not able to develop effective HCV-specific CD4$^+$ T-cell responses and have alterations in the cytokine production [67–69]. It has also been shown that T cells from persons with schistosomiasis have reduced IFN-γ–, IL-4–, and IL-10–secreting HCV-specific T-cell responses [69]. Elrefaei et al. characterized the HCV-specific CD8$^+$ T-cell responses in a cohort of HCV-infected individuals with or without *S. mansoni* coinfection.

Effect of Helminths on Acquisition of HCV and HIV Infections

In vitro data support the hypothesis that cells from helminth-infected people are more susceptible to infection with HIV-1. However, a more important public health question is whether a similar effect occurs in vivo. Many investigators have shown that cells from persons with schistosomiasis, intestinal helminths, or filariasis are more susceptible to HIV-1 infection than are cells from persons without helminthic infection [71–73]. The increase in the level of chemokine receptors that serve as HIV-1 coreceptors may provide a mechanistic explanation for these observations [70–74]. Secor et al. observed that CD4+ T cells and monocytes from schistosomiasis patients express higher levels of CXCR4 and CCR5 than do cells from patients who have previously had schistosomiasis but have been treated [75]. The increased density of CXCR4 and CCR5 on the surfaces of cells from patients with helminthiasis is probably because of the result of upregulation by the Th2-associated cytokines IL-4 and IL-10 [76]. Another factor that may increase the susceptibility of helminth-infected people to HIV is the upregulation of CTLA-4 in CD4+ T cells of patients with chronic helminthiases [77].

In HIV-infected persons, CTLA-4 engagement counteracts the CD28 antiviral effects, and the ratio of CTLA-4 to CD28 engagement determines the susceptibility of HIV-1 infection. Furthermore, unopposed CTLA-4 signaling provided by CD28 blockade promotes vigorous HIV-1 replication, despite minimal T-cell proliferation. Finally, anti–CTLA-4 antibodies decrease the susceptibility to HIV of antigen-activated CD4 T cells, suggesting a potential approach to prevent or limit viral spread in HIV-1–infected individuals [78]. The reduction of HIV-1 coreceptor densities on the surfaces of CD4+ T cells from HIV-1–positive schistosomiasis patients after praziquantel treatment may provide a beneficial effect in terms of reducing the susceptibility of the patient to the virus at exposure [75]. Similarly, treatment of filariasis patients reduced the susceptibility of their cells to infection with HIV-1 in vitro [73]. Together, the above data and those suggesting that Th2 cells, dominant in helminthic infections, are more readily infected with HIV [79] provide clues for the increased susceptibility to HIV of cells of patients with helminth infection.

Effect of Helminths on Concomitant HIV/HCV Progression

The interaction between HIV and the cytokine milieu of the host is believed to determine the course of the disease. The possible protective and beneficial role of Th1 and cytotoxic T lymphocyte (CTL) on the progression of HIV infection has been established [79,80]. Very large numbers of people with HIV are infected with chronic helminthic infections, which are characterized by a biased Th2 response and downregulated Th1 and CTL activity, thus raising the possibility of accelerated progression of HIV infection in people coinfected with helminths and the virus [81]. However, thusfar there is no sufficient evidence to prove this possibility [17,82]. Because it is not ethical to follow helminth-infected individuals for HIV-1 progression without treatment, human studies addressing this issue relied on measuring viral load (VL) at single time points before and after treatment. These studies produced controversial results. Some investigators found that deworming was associated with a VL reduction [81,83], whereas others did not find an overall association between treatment of helminthic infections and reduction in VL in coinfected adults [84,85]. Wolday et al. investigated the effect of anthelminthic treatment on HIV plasma VL in HIV- and helminth-infected individuals living in Ethiopia [81]. They found that, at baseline, HIV plasma VL was strongly correlated to the number of eggs excreted and was higher in individuals infected with more than one helminth. Successful deworming was associated with a significant decrease in HIV plasma VL124. Kallestrup et al. found that early praziquantel treatment of HIV–schistosomiasis coinfected Zimbabwean patients resulted in significant lower increase in VL and increase in CD4+ [83]. These results were confirmed experimentally by Chenine et al., who found that in a primate model of acute and chronic simian–human immunodeficiency virus, *S. mansoni* coinfection increased the expression of Th2-associated cytokine responses and virus replication, raising the possibility that parasitic infection may upregulate HIV replication in coinfected individuals [86]. Such an effect may of course affect the rate of progression of the HIV infection. Consequently, if helminth infections promote increased plasma VLs, this could in turn result in greater shedding of infectious virus into mucosal secretions, thus increasing the transmissibility of HIV-1. Indeed, persons with increased VLs are more likely to transmit virus to their sexual partners and may significantly contribute to the rapid spread of HIV-1 [87].

Conversely, many studies showed that helminthic infection in HIV-1 coinfected Ugandan adults was not associated with higher VL, lower CD4+, or faster disease progression [57,88,89]. They also found that effective anthelminthic treatment even resulted in transient increase of VL and decrease of CD4+ [57,84,88,89] and concluded that anthelminthic therapy may not be beneficial in slowing HIV progression in coinfected adults. The mechanism underlying the increase in VL after treatment is not clear; theoretically, treatment of schistosomes and nematode could release parasite antigen, causing Th2 lymphocyte activation, leading to increased HIV replication. Anthelminthic treatment might also eliminate parasite-induced production of immunosuppressive cytokines such as IL-10 and TGF-β [90]. An increase in type 2 cytokines occurring immediately after deworming was recorded by many workers [58,59]. Brown et al. found that treatment of *S. mansoni* in HIV coinfected patients enhanced schistosome-specific type 2 responses without increase in eosinophils or *M. tuberculosis*-specific IFN-γ response [57]. They also observed that treatment resulted in a significant decline in circulating IL-10 concentrations. So, it seems that treatment causes disruption of the regulatory network induced by chronic helminthiasis. As treatment of helminths is undoubtedly beneficial and recommended by international health organizations, the identification of new interventions for optimal eradication of infectious diseases, although leaving immunoregulatory mechanisms unchanged, would be very helpful [35].

Another aspect of public health importance is the impact of helminths on the acquisition of other pathogens by HIV-infected persons. Given that Th2 cytokines inhibit Th1 responses, it has been proposed that helminthic infections may enhance TB infection, whether newly acquired or reactivated. In an area with a high incidence of TB, such as South Africa, Beyers and coinvestigators found that serum IgE levels, as a marker of prominent Th2 responses, correlated with TB incidence and that total IgE and *Ascaris*-specific IgE levels were both high in the TB patients and declined after successful treatment [91,92]. Studies carried out in Uganda by Elliott et al. revealed profoundly impaired IFN-γ responses to mycobacterial antigens in association with HIV infection [93]. They also found a significant correlation between eosinophilia and increased incidence of TB in HIV$^+$ patients [94]. As eosinophilia is most likely indicative of type 2 response induced by helminth infection in this Ugandan cohort [94], it might be assumed that helminthic infections increase susceptibility to active TB in people coinfected with HIV. However, the mechanism of the observed association between eosinophilia and risk of TB remains to be determined [94].

CONCLUDING REMARKS

Helminthic infections are widespread all over the world. The interaction between helminthes and the host's immune system evokes particular immunomodulatory and immunoregulatory mechanisms that ensure their survival in the host for years. Chronic helminthic infections are characterized by immune activation together with biased Th2 response and downregulated Th1 and CTL activity. These changes in the immunologic milieu of the host might impair the immunologic response to bystander bacterial, viral, and protozoal pathogens, which mostly need Th1 responses to limit the severity and progression of infection. It might also have a big impact on the current and prospective vaccination programs. Therefore, from a global health perspective, there is a need to raise awareness at the national level to the consequences of interactions of multiple infections, particularly in the developing regions of the world. These efforts need to be complemented by political commitment and funding at the national and international level for research and disease control and prevention programs for infectious diseases.

REFERENCES

[1] Lucey DR, Clerici M, Shearer GM. Type 1 and type 2 cytokine dysregulation in human infectious, neoplastic, and inflammatory diseases. Clin Microbiol Rev 1996;9:542–62.

[2] Maizels RM, Yazdanbakhsh M. Immune regulation by helminth parasites: cellular and molecular mechanisms. Nat Rev Immunol 2003;3:733–44.

[3] Infante-Duarte C, Kamradt T. Th1/Th2 balance in infection. Springer Semin Immunopathol 1999;21:317–38.

[4] Loukas A, Prociv P. Immune responses in hookworm infections. Clin Microbiol Rev 2001;14:689–703.

[5] Yazdanbakhsh M, van den Biggelaar A, Maizels RM. Th2 responses without atopy: immunoregulation in chronic helminth infections and reduced allergic disease. Trends Immunol 2001;22:372–7.

[6] Macdonald AS, Loke P, Martynoga R, Dransfield I, Allen JE. Cytokine-dependent inflammatory cell recruitment patterns in the peritoneal cavity of mice exposed to the parasitic nematode. Brugia Malayi Med Microbiol Immunol 2003;192:33–40.

[7] Boitelle A, Scales HE, Di Lorenzo C, et al. Investigating the impact of helminth products on immune responsiveness using a TCR transgenic adoptive transfer system. J Immunol 2003;171:447–54.

[8] Coffman RL, von der Weid T. Multiple pathways for the initiation of T helper 2 (Th2) responses. J Exp Med 1997;185:373–5.

[9] O'Garra A. Commitment factors for T helper cells. Curr Biol 2000;10:R492–4.

[10] Rincón M, Anguita J, Nakamura T, Fikrig E, Flavell RA. IL-6 directs the differentiation of IL-4-producing CD4+ T cells. J Exp Med 1997;185:461–9.

[11] Tawill S, Le Goff L, Ali F, Blaxter M, Allen JE. Both free-living and parasitic nematodes induce a characteristic Th2 response that is dependent on the presence of intact glycans. Infect Immun 2004;72:398–407.

[12] Mulcahy G, O'Neill S, Donnelly S, Dalton JP. Helminths at mucosal barriers interaction with the immune system. Adv Drug Deliv Rev 2004;56:853–68.

[13] Del Pozo V, De Andres B, Martin E, et al. Eosinophil as antigen-presenting cell: activation of T cell clones and T cell hybridoma by eosinophils after antigen processing. Eur J Immunol 1992;22:1919–25.

[14] Sabin EA, Araujo MI, Carvalho EM, Pearce EJ. Impairment of tetanus toxoid-specific Th1-like immune responses in humans infected with *Schistosoma mansoni*. J Infect Dis 1996;173:269–72.

[15] Hida S, Tadachi M, Saito T, Taki S. Negative control of basophil expansion by IRF-2 critical for the regulation of Th1/Th2 balance. Blood 2005;106:2011–7.

[16] Borkow G, Leng Q, Weisman Z, et al. Chronic immune activation associated with intestinal helminth infections results in impaired signal transduction and anergy. J Clin Invest 2000;106:1053–60.

[17] Borkow G, Bentwich Z. Chronic immune activation associated with chronic helminthic and human immunodeficiency virus infections: role of hyporesponsiveness and anergy. Clin Microbiol Rev 2004;17:1012–30.

[18] King CL. Transmission intensity and human immune responses to lymphatic filariasis. Parasite Immunol 2001;23:363–71.

[19] Nutman TB, Kumaraswami V. Regulation of the immune response in lymphatic filariasis: perspectives on acute and chronic infection with *Wuchereria bancrofti* in South India. Parasite Immunol 2001;23:389–99.

[20] Yazdanbakhsh M. Common features of T cell reactivity in persistent helminth infections: lymphatic filariasis and schistosomiasis. Immunol Lett 1999;65:109–15.

[21] Gallin M, Edmonds K, Ellner JJ, et al. Cell-mediated immune responses in human infection with *Onchocerca volvulus*. J Immunol 1988;140:1999–2007.

[22] Patriarca PA, Wright PF, John TJ. Factors affecting the immunogenicity of oral poliovirus vaccine in developing countries: review. Rev Infect Dis 1991;13:926–39.

[23] Muniz-Junqueira MI, Tavares-Neto J, Prata A, Tosta CE. Antibody response to *Salmonella typhi* in human schistosomiasis mansoni. Rev Soc Bras Med Trop 1996;29:441–5.

[24] Haseeb MA, Craig JP. Suppression of the immune response to diphtheria toxoid in murine schistosomiasis. Vaccine 1997;15:45–50.

[25] Elias D, Wolday D, Akuffo H, Petros B, Bronner U, Britton S. Effect of deworming on human T cell responses to mycobacterial antigens in helminth-exposed individuals before and after bacille Calmette-Guérin (BCG) vaccination. Clin Exp Immunol 2001;123:219–25.

[26] Cooper PJ, Chico M, Sandoval C, et al. Human infection with *Ascaris lumbricoides* is associated with suppression of the interleukin-2 response to recombinant cholera toxin B subunit following vaccination with the live oral cholera vaccine CVD 103-HgR. Infect Immun 2001;69:1574–80.

[27] Loke P, Macdonald AS, Robb A, Maizels RM, Allen JE. Alternatively activated macrophages induced by nematode infection inhibit proliferation via cell-to-cell contact. Eur J Immunol 2000;30:2669–78.

[28] Elliott DE, Summers RW, Weinstock JV. Helminths and the modulation of mucosal inflammation. Curr Opin Gastroenterol 2005;21:51–8.

[29] Atochina O, Daly-Engel T, Piskorska D, McGuire E, Harn DAA. Schistosome-expressed immunomodulatory glycoconjugate expands peritoneal Gr1($^+$) macrophages that suppress naive CD4($^+$) T cell proliferation via an IFN-gamma and nitric oxide-dependent mechanism. J Immunol 2001;167:4293–302.

[30] Smith P, Walsh CM, Mangan NE, et al. Regulation of programmed death ligand 1 on macrophages. J Immunol 2004;173:1240–8.

[31] Knopf PM. Immunomodulation and allergy. Allergy Asthma Proc 2000;21:215–20.

[32] Araujo MI, Lopes AA, Medeiros M, et al. Inverse association between skin response to aeroallergens and *Schistosoma mansoni* infection. Int Arch Allergy Immunol 2000;123:145–8.

[33] Lynch NR, Hagel IA, Palenque ME, et al. Relationship between helminth infection and IgE response in atopic and nonatopic children in a tropical environment. J Allergy Clin Immunol 1998;101:217–21.

[34] Lynch NR, Hagel IA, Perez M, Di Prisco MC, Lopez R, Alvarez N. Effect of anthelminthic treatment on the allergic reactivity of children in a tropical slum. J Allergy Clin Immunol 1993;92:404–11.

[35] van den Biggelaar AH, Rodrigues LC, van Ree R, et al. Long-term treatment of intestinal helminths increases mite skin-test reactivity in Gabonese school children. J Infect Dis 2004;189:892–900.

[36] Palmer J, Celedón JCM, Weiss ST, Wang B, Fang Z, Xu X. *Ascaris lumbricoides* infection is associated with increased risk of childhood asthma and atopy in rural China. Am J Respir Crit Care Med 2002;165:1489–93.

[37] Lynch NR, Palenque M, Hagel I, Di Prisco MC. Clinical improvement of asthma after anthelminthic treatment in a tropical situation. Am J Respir Crit Care Med 1997;156:50–4.

[38] Hagel I, Lynch NR, Perez M, Di Prisco MC, Lopez R, Rojas E. Modulation of the allergic reactivity of slum children by helminth infection. Parasite Immunol 1993;15:311–5.

[39] Godfrey RC, Gradidge CF. Allergic sensitisation of human lung fragments prevented by saturation of IgE binding sites. Nature 1976:484–6.

[40] Cooper PJ, Mancero T, Espinel M, et al. Early human infection with *Onchocerca volvulus* is associated with an enhanced parasite-specific cellular immune response. J Infect Dis 2001;183:1662–8.

[41] Paterson JC, Garside P, Kennedy MW, Lawrence CE. Modulation of a heterologous immune response by the products of *Ascaris suum*. Infect Immun 2002;70:6058–67.

[42] Flores-Villanueva PO, Zheng XX, Strom TB, Stadecker MJ. Recombinant IL-10 and IL-10/Fc treatment down-regulate egg antigen-specific delayed hypersensitivity reactions and egg granuloma formation in schistosomiasis. J Immunol 1996;156:3315–20.

[43] Hoffmann KF, Cheever AW, Wynn TA. IL-10 and the dangers of immune polarization: excessive type 1 and type 2 cytokine responses induce distinct forms of lethal immunopathology in murine schistosomiasis. J Immunol 2000;164:6406–16.

[44] Sher A, Gazzinelli RT, Oswald IP, et al. Role of T-cell derived cytokines in the downregulation of immune responses in parasitic and retroviral infection. Immunol Rev 1992;127:183–204.

[45] Doetze A, Satoguina J, Burchard G, et al. Antigen-specific cellular hyporesponsiveness in a chronic human helminth infection is growth factor-beta but not by a T(h)1 to T(h)2 shift. Int Immunol 2000;12:623–30.

[46] Weiner HL. Induction and mechanism of action of transform- ing growth factor-beta secreting Th3 regulatory cells. Immunol Rev 2001;182:207–14.

[47] Hoerauf A, Brattig N. Resistance and susceptibility in human onchocerciasis-beyond Th1 vs. Th2. Trends Parasitol 2002;18:25–31.

[48] Wilson MS, Taylor MD, Balic A, Finney CAM, Lamb JR, Maizels RM. Suppression of allergic airway inflammation by helminth-induced regulatory T cells. J Exp Med 2005;202:1199–212.

[49] Fontenot JD, Gavin MA, Rudensky AY. Foxp3 programs the development and function of CD4+CD25+ regulatory T cells. Nat Immunol 2003;4:330–6.

[50] Moore KW, de Waal Malefyt R, Coffman RL, O'Garra A. Interleukin-10 and the interleukin-10 receptor. Annu Rev Immunol 2001;19:683–765.

[51] Elias D, Akuffo H, Pawlowski A, Haile M, Schon T, Britton S. *Schistosoma mansoni* infection reduces the protective efficacy of BCG vaccination against virulent *Mycobacterium tuberculosis*. Vaccine 2005;23:1326–34.

[52] Elias D, Akuffo H, Thors C, Pawlowski A, Britton S. Low dose chronic *Schistosoma mansoni* infection increases susceptibility to *Mycobacterium bovis* BCG infection in mice. Clin Exp Immunol 2005;139:398–404.

[53] Rougemont A, Boisson-Pontal ME, Pontal PG, Gridel F, Sangare S. Tuberculin skin tests and B.C.G. vaccination in hyperendemic area of onchocerciasis. Lancet 1977;1:309.

[54] Malhotra I, Mungai P, Wamachi A, et al. Helminth- and Bacillus Calmette-Guerin-induced immunity in children sensitized in utero to filariasis and schistosomiasis. J Immunol 1999;162:6843–8.

[55] Cooper PJ, Espinel I, Paredes W, Guderian RH, Nutman TB. Impaired tetanus-specific cellular and humoral responses following tetanus vaccination in human onchocerciasis: a possible role for interleukin-10. J Infect Dis 1998;178:1133–8.

[56] Borkow G, Bentwich Z. Eradication of helminthic infections may be essential for successful vaccination against HIV and tuberculosis. Bull WHO 2000;78:1368–9.

[57] Brown M, Mawa PA, Joseph S, et al. Treatment of *Schistosoma mansoni* infection increases helminth-specific type 2 cytokine responses and HIV-1 loads in coinfected Ugandan adults. J Infect Dis 2005;191:1648–57.

[58] Joseph S, Jones FM, Walter K, et al. Increases in human T helper 2 cytokine responses to *Schistosoma mansoni* worm and worm tegument antigens are specifically induced by treatment with praziquantel. J Infect Dis 2004;190:835842.

[59] Satti M, Cahen P, Skov P, et al. Changes in IgE- and antigen-dependent histamine-release in peripheral blood of *Schistosoma mansoni*-infected Ugandan fishermen after treatment with praziquantel. BMC Immunol 2004;5:6.

[60] Elliott AM, Namujju PB, Mawa PA, et al. Mother and Baby study team. A randomised controlled trial of the effects of albendazole in pregnancy on maternal responses to mycobacterial antigens and infant responses to bacille Calmette-Guérin (BCG) immunization. BMC Infect Dis 2005;5:115.

[61] Gupta M, Arora KL, Mithal S, Tandon BN. Effect of periodic deworming on nutritional status of *Ascaris*-infested preschool children receiving supplementary food. Lancet 1977;2:108–10.

[62] Bundy DA, de Silva NR. Can we deworm this wormy world? Br Med Bull 1998;54:421–32.

[63] Allen HE, Crompton DWT, de Silva N, Loverde PT, Olds GR. New policies for using anthelminthics in high risk groups. Trends Parasitol 2002;18:381–2.

[64] Beasley NM, Tomkins AM, Hall A, et al. The impact of population level deworming on the haemoglobin levels of school children in Tanga, Tanzania. Trop Med Int Health 1999;4:744–50.

[65] Actor JK, Shirai M, Kullberg MC, Buller RM, Sher A, Berzofsky JA. Helminth infection results in decreased virus- specific CD8+ cytotoxic T-cell and Th1 cytokine responses as well as delayed virus clearance. Proc Natl Acad Sci USA 1993;90:948–52.

[66] Edwards MJ, Buchatska O, Ashton M, Montoya M, Bickle QD, Borrow P. Reciprocal immunomodulation in a schistosome and hepatotropic virus coinfection model. J Immunol 2005;175:6275–85.

[67] Kamal SM, Rasenack JW, Bianchi L, et al. Acute hepatitis C without and with schistosomiasis: correlation with hepatitis C-specific CD4+ T-cell and cytokine response. Gastroenterology 2001;121:646–56.

[68] Kamal SM, Bianchi L, Al Tawil A, et al. Specific cellular immune response and cytokine patterns in patients coinfected with hepatitis C virus and *Schistosoma mansoni*. J Infect Dis 2001;184:972–82.

[69] Farid A, Al-Sherbiny M, Osman A, et al. *Schistosoma* infection inhibits cellular immune responses to core HCV peptides. Parasite Immunol 2005;27:189–96.

[70] Marlink R, Kanki P, Thior I, et al. Reduced rate of disease development after HIV-2 infection as compared to HIV-1. Science 1994;265:1587–90.

[71] Shapira-Nahor O, Kalinkovich A, Weisman Z, et al. Increased susceptibility to HIV-1 infection of peripheral blood mononuclear cells from chronically immune-activated individuals. AIDS 1998;12:1731–3.

[72] Kalinkovich A, Weisman Z, Leng Q, et al. Increased CCR5 expression with decreased β-chemokine secretion in Ethiopians: relevance to AIDS in Africa. J Hum Virol 1999;2:283–9.

[73] Gopinath R, Ostrowski M, Justement SJ, Fauci AS, Nutman TB. Filarial infections increase susceptibility to human immunodeficiency virus infection in peripheral blood mononuclear cells in vitro. J Infect Dis 2000;182:1804–8.

[74] Moonis M, Lee B, Bailer RT, Luo Q, Montaner LJ. CCR5 and CXCR4 expression correlated with X4 and R5 HIV-1 infection yet not sustained replication in Th1 and Th2 cells. AIDS 2001;15:1941–9.

[75] Secor WE, Shah A, Mwinzi PM, Ndenga BA, Watta CO, Karanja DM. Increased density of human immunodeficiency virus type 1 coreceptors CCR5 and CXCR4 on the surfaces of CD4+ T cells and monocytes of patients with *Schistosoma mansoni* infection. Infect Immun 2003;71:6668–71.

[76] Wang J, Crawford K, Yuan M, Wang H, Gorry PR, Gabuzda D. Regulation of CC chemokine receptor 5 and CD4 expression and human immunodeficiency virus type 1 replication in human macrophage and microglia by T helper type 2 cytokines. J Exp Med 2002;185:885–97.

[77] Steel C, Nutman TB. CTLA-4 in filarial infections: implications for a role in diminished T cell reactivity. J Immunol 2003;170:1930–8.

[78] Riley JL, Schlienger K, Blair BJ, et al. Modulation of susceptibility to HIV-1 infection by the cytotoxic T lymphocyte antigen 4 costimulatory molecule. J Exp Med 2000;191:1987–97.

[79] Maggi E, Mazzetti M, Ravina A, et al. Ability of HIV to promote a TH1 to TH0 shift and to replicate preferentially in TH2 and TH0 cells. Science 1994;265:244–8.

[80] Clerici M, Shearer GM. The TH1-TH2 hypothesis of HIV infection: new insights. Immunol Today 2001;12:575–81.

[81] Wolday D, Mayaan S, Mariam ZG, et al. Treatment of intestinal worms is associated with decreased HIV plasma viral load. J Acquir Immune Defic Syndr 2002;31:56–62.

[82] Secor WE, Karanja DMS, Colley DG. Interactions between schistosomiasis and human immunodeficiency virus in Western Kenya. Mem Inst Oswaldo Cruz 2004;99(Suppl. I): 93–5.

[83] Kallestrup P, Zinyama R, Gomo E, et al. Schistosomiasis and HIV-1 infection in rural Zimbabwe: effect of treatment of schistosomiasis on CD4 cell count and plasma HIV-1 RNA load. J Infect Dis 2005;192:1956–61.

[84] Lawn SD, Karanja DM, Mwinzia P, et al. The effect of treatment of schistosomiasis on blood plasma HIV-1 RNA concentration in coinfected individuals. AIDS 2000;14:2437–43.

[85] Modjarrad K, Zulu I, Redden DT, et al. Treatment of intestinal helminths does not reduce plasma concentrations of HIV-1 RNA in coinfected Zambian adults. J Infect Dis 2005;192:1277–83.

[86] Chenine AL, Buckley KA, Li PL, et al. *Schistosoma mansoni* infection promotes SHIV clade C replication in rhesus macaques. AIDS 2005;19:1793–7.

[87] Quinn TC, Wawer MJ, Sewankambo N, et al. Viral load and heterosexual transmission of human immunodeficiency virus type 1. Rakai Project Study Group. N Engl J Med 2000;342:921–9.

[88] Elliott AM, Mawa PA, Joseph S, et al. Associations between helminth infection and CD4+ T cell count, viral load and cytokine responses in HIV-1infected Ugandan adults. Trans R Soc Trop Med Hyg 2003;97:103–8.

[89] Brown M, Kizza M, Watera C, et al. Helminth infection is not associated with faster progression of HIV disease in coinfected adults in Uganda. J Infect Dis 2004;190:1869–79.

[90] Morgan D, Whitworth J. The natural history of HIV-1 infection in Africa. Nat Med 2001;7:143–5. 128.

[91] Beyers AD, Van Rie A, Adams J, Fenhalls G, Gie R, Beyers N. Signals that regulate the host response to *Mycobacterium tuberculosis*. Novartis Found Symp 1998;217:145–57.

[92] Adams JF, Scholvinck EH, Gie RP, Potter PC, Beyers N, Beyers AD. Decline in total serum IgE after treatment for tuberculosis. Lancet 1999;353:2030–3.

[93] Elliott AM, Hurst TJ, Balyeku MN, et al. The immune response to *Mycobacterium tuberculosis* in HIV-infected and uninfected adults in Uganda: application of a whole blood cytokine assay in an epidemiological study. Int J Tuberc Lung Dis 1999;3:239–47.

[94] Elliott AM, Kyosiimire J, Quigley MA, et al. Eosinophilia and progression to active tuberculosis in HIV-1-infected Ugandans. Trans R Soc Trop Med Hyg 2003;97:477–80.

Chapter 4.3

Hepatitis C and HIV Coinfection in Developing Countries

Ozlem Tastan Bishop
Rhodes University, Grahamstown, South Africa

Chapter Outline

HIV IN DEVELOPING COUNTRIES

The United Nations currently estimates that worldwide, there are 36.7 million people living with HIV, and 2.1 million (1.8 million–2.4 million) people became newly infected with HIV (Tables 4.3.1 and 4.3.2) [1,2]. The vast majority of HIV-infected people, approximately 95% of the total, live in low- and middle-income countries [2,3] (Fig. 4.3.1). Sub-Saharan Africa has been hit especially hard, with almost 70% of all HIV-infected patients living there and 90% of all maternal–fetal transmission occurring there. Although the adult prevalence rate in most developed countries is less than 1%, it is 8% in sub-Saharan Africa. In most of those countries, poverty and a lack of resources make caring for HIV-infected patients and effective prevention efforts virtually impossible [1,2,4]. Even in countries with moderate resources (for example, India), HIV infection is spreading rapidly among at-risk populations, especially intravenous drug users [2,3].

Hepatitis C in Developing Countries. https://doi.org/10.1016/B978-0-12-803233-6.00012-6
135

TABLE 4.3.1 Global HIV Data 2016 [1]

Population	2000	2005	2010	2011	2012	2013	2014	2015/2016
People living with HIV	28.9 M [26.5–31.7 M]	31.8 M [29.4–34.5 M]	33.3 M [30.8–36.1 M]	33.9 M [31.4–36.7 M]	34.5 M [31.9–37.4 M]	35.2 M [32.6–38.1 M]	35.9 M [33.3–38.9 M]	36.7 M [34.0–39.8 M]
New HIV infections (total)	3.2 M [2.9–3.5 M]	2.5 M [2.3–2.8 M]	2.2 M [2.0–2.5 M]	2.2 M [1.9–2.5 M]	2.2 M [1.9–2.4 M]	2.1 M [1.9–2.4 M]	2.1 M [1.9–2.4 M]	2.1 M [1.8–2.4 M]
New HIV infections (aged ≥15)	2.7 M [2.5–3.0 M]	2.1 M [1.9–2.3 M]	1.9 M [1.7–2.1 M]	1.9 M [1.7–2.2 M]	1.9 M [1.7–2.2 M]	1.9 M [1.7–2.2 M]	1.9 M [1.7–2.2 M]	1.9 M [1.7–2.2 M]
New infections (aged 0–14)	490,000 [430,000–560,000]	450,000 [390,000–510,000]	290,000 [250,000–350,000]	270,000 [220,000–330,000]	230,000 [190,000–290,000]	200,000 [160,000–250,000]	160,000 [130,000–220,000]	150,000 [110,000–190,000]
AIDS-related deaths	1.5 M [1.3–1.8 M]	2.0 M [1.7–2.3 M]	1.5 M [1.3–1.7 M]	1.4 M [1.2–1.7 M]	1.4 M [1.2–1.6 M]	1.3 M [1.1–1.5 M]	1.2 M [990,000–1.4 M]	1.1 M [940,000–1.3 M]
People accessing treatment	770,000 [680,000–800,000]	2.2 M [1.9–2.2 M]	7.5 M [6.6–7.8 M]	9.1 M [8.0–9.5 M]	11 M [9.6–11.4 M]	13 M [11.4–13.5 M]	15 M [13.2–15.6 M]	18.2 M [16.1–19.0 M] (*June 2016) 17 M [15.0–17 M] (end 2015)
Resources available for HIV (low- and middle-income countries)	4.8 billion	9.4 billion	15.9 billion	18.3 billion	19.5 billion	19.6 billion	19.2 billion	19 billion

TABLE 4.3.2 Regional HIV Data 2015 [1,2]

Region	People Living With HIV (Total)	New HIV Infections (Total)	New HIV Infections (Aged 15+)	New HIV Infections (Aged 0–14)	AIDS-Related Deaths (Total)	Total Number Accessing Antiretroviral Therapy
Eastern and southern Africa	19.0 million [17.7–20.5 million]	960,000 [830,000–1.1 million]	910,000 [790,000–1.1 million]	56,000 [40,000–76,000]	470,000 [390,000–560,000]	10 million
Latin America and the Caribbean	2.0 million [1.7–2.3 million]	100,000 [86,000–120,000]	100,000 [84,000–120,000]	2100 [1600–2900]	50,000 [41,000–59,000]	1.1 million
Western and central Africa	6.5 million [5.3–7.8 million]	410,000 [310,000–530,000]	350,000 [270,000–450,000]	66,000 [47,000–87,000]	330,000 [250,000–430,000]	1.8 million
Asia and the Pacific	5.1 million [4.4–5.9 million]	300,000 [240,000–380,000]	280,000 [220,000–350,000]	19,000 [16,000–21,000]	180,000 [150,000–220,000]	2.1 million
Eastern Europe and central Asia	1.5 million [1.4–1.7 million]	190,000 [170,000–200,000]	190,000 [170,000–200,000]	---*	47,000 [39,000–55,000]	320,000
Middle East and North Africa	230,000 [160,000–330,000]	21,000 [12,000–37,000]	19,000 [11,000–34,000]	2100 [1400–3200]	12,000 [8700–16,000]	38,000
Western and central Europe and North America	2.4 million [2.2–2.7 million]	91,000 [89,000–97,000]	91,000 [88,000–96,000]	---*	22,000 [20,000–24,000]	1.4 million

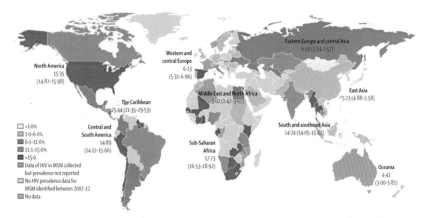

FIGURE 4.3.1 Worldwide prevalence of HIV. *(Beyrer C, Baral S, van Griensven F, Goodreau SM, Chariyalertsak S, Wirtz A, Brookmeyer R. Global epidemiology of HIV infection in men who have sex with men. The Lancet July 28, 2012;380.)*

An estimated 25.5 million people living with HIV live in sub-Saharan Africa. The vast majority of them (an estimated 19 million) live in east and southern Africa. Around 40% of all people living with HIV do not know that they have the virus [4–6]. Developing nations, especially those in sub-Saharan Africa and the Caribbean basin, continue to lack the resources to provide even the most basic prophylactic medications or health care to HIV-infected populations.

HIV EPIDEMIC IN SUB-SAHARAN AFRICA

Sub-Saharan Africa contains 12% of the global population; however, it accounts for more than 71% of the global burden of HIV infection. Ten African countries, namely, South Africa (25%), Nigeria (13%), Mozambique (6%), Uganda (6%), Tanzania (6%), Zambia (4%), Zimbabwe (6%), Kenya (6%), Malawi (4%), and Ethiopia (3%), account for almost 80% of all people living with HIV [1,6,7].

New HIV infections across countries in sub-Saharan Africa have shown a decline by more than 33% from an estimated 2.2 (2.1–2.3) million in 2005 to 1.5 (1.3–1.6) million in 2013, but still remain high. The widespread coverage of ART has led to substantial declines in new HIV infections. For example, HIV-uninfected individuals living in a community with high ART coverage (30%–40% of all HIV-infected individuals on ART) were 38% less likely to acquire HIV than those living in communities where ART coverage was low (<10% of all HIV-infected individuals on ART) [8]. Despite these declines, HIV incidence rates remain unacceptably high with the largest number of new infections coming from South Africa (23%), Nigeria (15%), Uganda (10%), Mozambique (8%), and Kenya (7%). Nevertheless, the epidemics in Botswana, Namibia, and

Zambia seem to be declining, whereas the epidemics in Lesotho, Mozambique, and Swaziland did not change [9].

In sub-Saharan Africa, the main mode of HIV transmission is through heterosexual sex with a concomitant epidemic in children through vertical transmission. As such, women are disproportionately affected accounting for 58% of the total number of people living with HIV and have the highest number of children living with HIV and the highest number of AIDS-related deaths [3,4].

With increasing access to ART, the number of AIDS-related deaths have steadily declined and in sub-Saharan Africa these decreased by 39% between 2005 and 2013 [9]. In South Africa the decline was 51%, whereas in Ethiopia it was 37% and in Kenya it was 32%. Several empiric studies from South Africa, Uganda, Tanzania, Rwanda, and Malawi have demonstrated that the impact of modest ART coverage at CD4 cell counts ranging from <200 to 500 cells/μL resulted in significant declines in mortality with life expectancy increasing by an additional 10 years. Despite these benefits of ART, in 2013 the region still accounted for 74% of deaths from AIDS-related illnesses [10–17].

HIV IN DEVELOPING COUNTRIES IN LATIN AMERICA AND THE CARIBBEAN

An estimated 2 million people are living with HIV in Latin America and the Caribbean combined, including 100,000 newly infected in 2015. Annual new HIV infections among adults increased by 2% in Latin America and by 9% in the Caribbean between 2010 and 2015. Nine countries in Latin America and the Caribbean have generalized epidemics. Of the countries with available data, The Bahamas has the region's highest prevalence (3.2%), and Brazil the greatest number of people living with the disease (830,000) [2,3].

HEPATITIS C AND HIV COINFECTION

An estimated 2.3 million people living with HIV are coinfected with hepatitis C virus (HCV) globally. More than half, or 1.3 million, are people who inject drugs (PWID). HIV-infected people are on average six times more likely than HIV-uninfected people. HIV and HCV infections are major global public health problems, with overlapping modes of transmission and affected populations. Globally, there are 37 million people infected with HIV, and around 115 million people with chronic HCV infection [18,19].

Chronic HCV infection has become a major threat to the survival of HIV-infected persons in areas where antiretroviral therapy (ART) is available [18]. Increasingly, incident HCV among HIV-infected men who have sex with men is associated with sexually risky behavior and further research should be performed to refine understanding of the causal mechanism of this association [19]. Transmission of HIV and HCV can occur through percutaneous exposure to blood, sexual intercourse, and from a mother to her infant, but the efficiency of transmission varies for each virus. HCV is ~10 times more infectious than HIV through percutaneous blood exposures [20], and HIV-infected PWID represent the majority of HCV/HIV coinfection. More specifically, HCV coinfection rates among HIV-infected individuals who use injection drugs often exceed 90%, as demonstrated in studies in Europe and Asia [18].

Studies have revealed that transmission of HIV through sexual intercourse and from a mother to her infant is more efficient than for HCV transmission, although the efficiency of HCV transmission in these settings might increase in the presence of coinfection [21]. For example, in the Women and Infants Transmission Study [22], the risk of HCV infection from mother to infant was 3.2-fold greater in HIV-1–infected infants compared with HIV-1–uninfected infants (17.1% vs. 5.4%), and a meta-analysis reported similar findings [23] (Table 4.3.3).

IMPACT OF HIV INFECTION ON CHRONIC HCV

Several studies showed that HIV infection is an independent risk factor for the presence of liver disease in HCV infection [21–27]. HCV and HIV coinfection is characterized by a greater degree of portal, periportal and lobular inflammation, and centrilobular fibrosis [23,25–27]. In patients with CD4$^+$ T-cell counts below 200 cells/μL, histologic evidence of cirrhosis has been demonstrated in almost 50% of patients compared with 10% of patients with higher CD4$^+$ T-cell counts [26]. These suggest that both initial and later stage hepatic injury are more common in coinfected individuals. The potential importance of other factors, such as ART, in explaining the prevalence of liver disease in coinfected individuals is supported by the fact that the three classic histologic lesions associated with HCV infection (i.e., steatosis, bile duct injury, and lymphoid nodules) were not more frequently identified in coinfected patients [23].

Evidence that HIV infection affects the rate of hepatic fibrosis has been reported [25]. In this study, fibrosis progression rates were calculated by dividing Metavir fibrosis stage values by the estimated years of HCV infection. The estimated median duration of HCV infection before cirrhosis was 26 years (range, 22–34 years) in coinfected patients versus 38 years for HIV-seronegative people (range, 32–47 years) and was statistically significant. Furthermore, this study provided additional evidence that impaired immunity (CD4 count less than 200 cells/μL) and alcohol consumption greater than 50 g/week were independently associated with rapid progression to cirrhosis. Sex was not found to be an independent predictor of fibrosis rate.

Several studies [28–30] have demonstrated that progression to clinically evident liver disease is also accelerated in coinfected individuals. In one study [28], 81 coinfected people with hemophilia infected between 1978 and 1985 were assessed for persistent bilirubin elevation greater than 21 μmol/L, ultrasound-demonstrated ascites, hepatitic encephalopathy, histologic evidence of cirrhosis or hepatocellular carcinoma, and esophageal or gastric varices. Alternative explanations for liver disease, including opportunistic infections, malignancies,

TABLE 4.3.3 HIV and HCV Prevalence and Incidence by Region, 2015 [2,3,18]

Region	HIV			HCV		
	Total No. Living With HIV N (100%)	Newly Infected	Adult Prevalence (%)	Total No. (%) Living With HCV	Newly Infected	Adult Prevalence (%)
Global Total	36.7 million (100%)	2.1 million	0.8	71 million	1.75 million	3.3%
Eastern and Southern Africa	19.0 Million (52%)	960,000	7.1	Eastern Africa 6.1 million — Southern Africa 1.4 million	NA	Eastern Africa 2.0 — Southern Africa 2.1
Central and Western Africa	6.5 million (18%)	410,000	2.2	Central sub-Saharan Africa 2.3 (1.6–3.1) 1.9 million — West sub-Saharan Africa 8.4 million	NA	Central sub-Saharan Africa 2.0 — West sub-Saharan Africa 2.8
Asia and the Pacific	5.1 million (14%)	300,000	0.2	10 million		0.5
Latin America and the Caribbean	2.0 million (5%)	100,000	0.5	7 million		1
North Africa and Middle East		21,000	0.1	Egypt 11 million — North Africa (other than Egypt) and Middle East 10 million		Egypt 14.7 — North Africa (other than Egypt) and Middle East 2.2

HCV, hepatitis C virus.

and alcohol, were ruled out before attributing these end points to HCV infection. The mean estimated time between HCV infection and any of these end points was 17.2 years. Twenty-two (27%) met at least one of these criteria compared with 5.7% in a matched HCV-infected, HIV-seronegative cohort. Based on this analysis, an odds ratio for progression to these clinical measures of liver disease was 7.4 (95% confidence interval 2.2 to 22.5). Unfortunately, no attempt was made to correlate antiretroviral use with these hepatic end points. Although the natural history of HCV infection appears to be accelerated in the presence of HIV infection, significant liver disease still requires many years to develop. Rapid deterioration in hepatic status after HIV infection in patients with pre-existing HCV infection has, however, been described as well [27].

EFFECT OF ART ON THE COURSE OF HCV

The first effective therapy against HIV was a nucleoside reverse transcriptase inhibitor (NRTI) followed by more effective three-drug therapy combining two NRTIs with a new class of antiretrovirals– protease inhibitors – and was incorporated into clinical practice. This was followed by a combination of at least three drugs (called "highly active," "combination" antiretroviral therapy, or HAART). The widespread use of antiretrovirals in developed countries has substantially reduced mortality in HIV-infected individuals. Given the longer life span of HIV-infected patients' after adoption of HAART, chronic HCV infection, which generally progresses to clinical disease over decades, has become an important cause of liver disease in coinfected individuals [31]. A cohort study of HIV-infected patients with hemophilia found that following the introduction of HAART, the proportion of deaths caused by AIDS had decreased, while the proportion of deaths caused by liver disease had increased [32].

ART is currently recommended in coinfected patients, including those with cirrhosis [33]. It has been suggested that the HIV viral suppression and immune reconstitution possible with HAART are critical factors that positively affect the rate of HCV fibrosis progression [34,35]. Some studies have reported that HAART may adversely affect hepatitis C outcomes by increasing HCV viral load, liver toxicity, and fibrosis progression [36–38]. It is possible that HAART can attenuate the progression of liver disease through the reversal or prevention of HIV-related immunosuppression, but it is also plausible that antiretroviral use may exacerbate liver disease [39]. The incidence of HAART-associated liver toxicity is approximately three times greater in HIV/HCV–co-infected individuals than in those without hepatitis C [25,40].

TREATMENT OF HCV WITH COEXISTING HIV INFECTION IN THE ERA OF DIRECT-ACTING ANTIVIRAL AGENTS [41–47]

HCV infection in patients with HIV infection can have significant consequences, including liver disease progression, cirrhosis, increased rates of end-stage liver disease, and shortened lifespan after hepatic decompensation [29,31]. Thus,

HCV treatment is recommended in patients coinfected with HIV. The peg-interferon and ribavirin combination regimen had particularly low SVR rates among those with genotype 1 HCV (typically less than 30%). Further, the SVR rates with peg-interferon and ribavirin was 15%–25% lower in coinfected patients than in HCV-monoinfected patients. In 2011, the addition of a first-generation HCV protease inhibitor (telaprevir or boceprevir) to peg-interferon and ribavirin improved SVR rates with GT1 to approximately 60% and narrowed the gap in treatment response between coinfected and monoinfected patients to approximately 15%. This triple-therapy regimen, however, proved to be quite complex and challenging because of interactions with antiretroviral medications, greater pill burden, food requirements, and additional adverse effects. With the advent of all-oral regimens that include new direct-acting antiviral agents (DAAs) that are safe, highly effective, and have fewer drug interactions, the older regimens of peg-interferon and ribavirin or peg-interferon and ribavirin plus a first-generation protease inhibitor are no longer recommended for treatment of HCV in coinfected patients [39,40]. However, access to the highly potent DAAs is considered a real challenge and an extra economic burden in resource-limited countries where both viral infections are frequent.

Recommended HCV Treatment in Patients With HIV Coinfection

ART should be considered for most patients with HIV and HCV coinfection, regardless of their CD4 cell count depending on the findings of several studies that showed that ART may slow liver disease progression and reduce the risk of liver-related morbidity. In antiretroviral-naive patients with a CD4 count greater than 500 cells/mm, ART could be deferred until after treatment of HCV. For those patients whose CD4 counts are less than 200 cells/mm, it may be advisable to first initiate ART and defer HCV therapy until the patient is stable on ART.

Factors to Consider before Choosing HCV Treatment Regimen: In persons coinfected with HIV, the major factors in selecting a regimen to treat HCV include HCV genotype, prior treatment experience, presence of cirrhosis, and potential drug interactions with the HIV antiretroviral medications. Of note, the preponderance of clinical trial data on the efficacy of HCV therapy derive from patients on ART with suppressed HIV RNA levels and CD4 counts greater than 200 cells/mm.

According to the AASLD and IDSA guidelines, patients with HIV and HCV coinfection are managed using the same approach for HCV treatment as those with HCV monoinfection, putting into consideration important drug–drug interactions between HCV medications and HIV antiretroviral medications. In the AASLD, IDSA, and EASL HCV guidance [41,42], older regimens, such as peg-interferon plus ribavirin or peg-interferon plus ribavirin plus either boceprevir, telaprevir, or simeprevir, are not recommended for use in any circumstance, because of suboptimal SVR rates, long duration of therapy, and poor tolerance. The recent recommendations put forth in 2016 for treating HCV in HIV-coinfected patients are summarized in Tables 4.3.4 and 4.3.5.

TABLE 4.3.4 Treatment of HCV in HIV Coinfected Patients [41,42]

Patients	Treatment-Naïve or -Experienced	Sofosbuvir/ Ledipasvir	Sofosbuvir/ Velpatasvir	Ombitasvir/ Paritaprevir/ Ritonavir and Dasabuvir	Ombitasvir/ Paritaprevir/ Ritonavir	Grazoprevir/Elbasvir	Sofosbuvir and Daclatasvir	Sofosbuvir and Simeprevir
G1a	Treatment-naïve	12 wk, no ribavirin	12 wk, no ribavirin	24 wk with ribavirin	No	12 wk, no ribavirin if HCV RNA ≤800,000 (5.9 log) IU/mL or 16 wk with ribavirin if HCV RNA >800,000 (5.9 log) IU/mL	12 wk, no ribavirin	No
	Treatment-experienced	12 wk with ribavirin or 24 wk, no ribavirin					12 wk with ribavirin[a] or 24 wk, no ribavirin	
G1b	Treatment-naïve	12 wk, no ribavirin	12 wk, no ribavirin	12 wk, no ribavirin	No	12 wk, no ribavirin	12 wk, no ribavirin	No
	Treatment-experienced							
G2	Both	No	12 wk, no ribavirin	No	No	No	12 wk, no ribavirin	No
G3	Treatment-naïve	No	12 wk with ribavirin or 24 wk, no ribavirin	No	No	No	24 wk with ribavirin	No
	Treatment-experienced							

G4	Treatment-naïve	12 wk, no ribavirin	12 wk, no ribavirin	No	12 wk with ribavirin	12 wk, no ribavirin	12 wk, no ribavirin	12 wk, no ribavirin
	Treatment-experienced	12 wk with ribavirin or 24 wk, no ribavirin	12 wk, no ribavirin	No	12 wk with ribavirin	12 wk, no ribavirin if HCV RNA ≤800,000 (5.9 log) IU/mL or 16 wk with ribavirin if HCV RNA >800,000 (5.9 log) IU/mL	12 wk with ribavirin or 24 wk, no ribavirin	12 wk with ribavirin or 24 wk, no ribavirin
G5 or 6	Treatment-naïve	12 wk, no ribavirin	12 wk, no ribavirin	No	No	No	12 wk, no ribavirin	No
	Treatment-experienced	12 wk with ribavirin or 24 wk, no ribavirin	12 wk, no ribavirin	No	No	No	12 wk with ribavirin or 24 wk, no ribavirin	No

Ribavirin is added only in patients with RASs that confer high-level resistance to NS5A inhibitors at baseline if RAS testing is available.
Therapy is prolonged to 16 weeks and ribavirin is added only in patients with RASs that confer resistance to elbasvir at baseline if RAS testing is available.
Ribavirin is added only in patients with NS5A RASs Y93H at baseline if RAS testing is available.
G, genotype; HCV, hepatitis C virus; RASs, relevant resistance-associated substitutions; wk, week.

TABLE 4.3.5 Treatment Recommendations for Retreatment of HCV-Monoinfected or HCV/ Antiviral Therapy Containing One or Several Direct-Acting Antiviral Agent(s)

Failed Treatment	Genotype	Sofosbuvir/ Ledipasvir	Sofosbuvir/ Velpatasvir	Ombitasvir/ Paritaprevir/ Ritonavir and Dasabuvir	Ombitasvir/ Paritaprevir/ Ritonavir	Grazoprevir/ Elbasvir
Peg-IFN-α with ribavirin and telaprevir, or boceprevir, or simeprevir	1	12 wk with ribavirin	12 wk with ribavirin	No	No	No
Sofosbuvir alone, or sofosbuvir plus ribavirin, or sofosbuvir plus Peg-IFN-α and ribavirin	1	12 wk with ribavirin (F0–F2) or 24 wk with ribavirin (F3–F4)	12 wk with ribavirin (F0–F2) or 24 wk with ribavirin (F3–F4)	12 wk with ribavirin (F0–F2) or 24 wk with ribavirin (F3–F4)	No	12 wk with ribavirin (F0–F2 with HCV RNA ≤800,000 (5.9 log) IU/mL) or 24 wk with ribavirin (F0–F2 with HCV RNA >800,000 (5.9 log) IU/mL and F3–F4)
	2	No	12 wk with ribavirin (F0–F2) or 24 wk with ribavirin (F3–F4)	No	No	No
	3	No	12 wk with ribavirin (F0–F2) or 24 wk with ribavirin (F3–F4)	No	No	No
	4	12 wk with ribavirin (F0–F2) or 24 wk with ribavirin (F3–F4)	12 wk with ribavirin (F0–F2) or 24 wk with ribavirin (F3–F4)	No	12 wk with ribavirin (F0-F2) or 24 wk with ribavirin (F3–F4)	12 wk with ribavirin (F0-F2 with HCV RNA ≤800,000 (5.9 log) IU/mL) or 24 wk with ribavirin (F0-F2 with HCV RNA >800,000 (5.9 log) IU/mL and F3–F4)
Failed treatment	Genotype	Sofosbuvir/ ledipasvir	Sofosbuvir/ velpatasvir	Ombitasvir/ paritaprevir/ ritonavir and dasabuvir	Ombitasvir/ paritaprevir/ ritonavir	Grazoprevir/ elbasvir

HIV-Coinfected Patients With Chronic Hepatitis C Who Failed to Achieve an SVR on Before

Sofosbuvir and Daclatasvir	Sofosbuvir and Simeprevir	Sofosbuvir Plus Ombitasvir/ Paritaprevir/ Ritonavir and Dasabuvir	Sofosbuvir Plus Ombitasvir/ Paritaprevir/ Ritonavir	Sofosbuvir Plus Grazoprevir/ Elbasvir	Sofosbuvir Plus Daclatasvir Plus Simeprevir
12 wk with ribavirin	No	No	No	No	No
12 wk with ribavirin (F0–F2) or 24 wk with ribavirin (F3–F4)	12 wk with ribavirin (F0–F2) or 24 wk with ribavirin (F3–F4)	No	No	No	No
12 wk with ribavirin (F0–F2) or 24 wk with ribavirin (F3–F4)	No	No	No	No	No
12 wk with ribavirin (F0–F2) or 24 wk with ribavirin (F3–F4)	No	No	No	No	No
12 wk with ribavirin (F0–F2) or 24 wk with ribavirin (F3–F4)	12 wk with ribavirin (F0–F2) or 24 wk with ribavirin (F3–F4)	No	No	No	No
Sofosbuvir and daclatasvir	Sofosbuvir and simeprevir	Sofosbuvir plus ombitasvir/ paritaprevir/ ritonavir and dasabuvir	Sofosbuvir plus ombitasvir/ paritaprevir/ ritonavir	Sofosbuvir plus grazoprevir/ elbasvir	Sofosbuvir plus daclatasvir plus simeprevir

Continued

TABLE 4.3.5 Treatment Recommendations for Retreatment of HCV-Monoinfected or HCV/
Antiviral Therapy Containing One or Several Direct-Acting Antiviral Agent(s)—cont'd

Failed Treatment	Genotype	Sofosbuvir/ Ledipasvir	Sofosbuvir/ Velpatasvir	Ombitasvir/ Paritaprevir/ Ritonavir and Dasabuvir	Ombitasvir/ Paritaprevir/ Ritonavir	Grazoprevir/ Elbasvir
Sofosbuvir alone, or sofosbuvir plus ribavirin, or sofosbuvir plus Peg-IFN-α and ribavirin	5 or 6	12 wk with ribavirin (F0–F2) or 24 wk with ribavirin (F3–F4)	12 wk with ribavirin (F0–F2) or 24 wk with ribavirin (F3–F4)	No	No	No
Sofosbuvir and simeprevir	1	12 wk with ribavirin (F0–F2) or 24 wk with ribavirin (F3–F4)	12 wk with ribavirin (F0–F2) or 24 wk with ribavirin (F3–F4)	No	No	No
	4	12 wk with ribavirin (F0–F2) or 24 wk with ribavirin (F3–F4)	12 wk with ribavirin (F0–F2) or 24 wk with ribavirin (F3–F4)	No	No	No
NS5A inhibitor-containing regimen (ledipasvir, velpatasvir, ombitasvir, elbasvir, daclatasvir)	1a	No	No	No	No	No
	1b	No	No	No	No	No
	2	No	24 wk with ribavirin	No	No	No
	3	No	24 wk with ribavirin	No	No	No
	4	No	No	No	No	No
	5 or 6	No	24 wk with ribavirin	No	No	No

HCV, hepatitis C virus; *Peg-IFN*, pegylated interferon.

HIV-Coinfected Patients With Chronic Hepatitis C Who Failed to Achieve an SVR on Before

Sofosbuvir and Daclatasvir	Sofosbuvir and Simeprevir	Sofosbuvir Plus Ombitasvir/ Paritaprevir/ Ritonavir and Dasabuvir	Sofosbuvir Plus Ombitasvir/ Paritaprevir/ Ritonavir	Sofosbuvir Plus Grazoprevir/ Elbasvir	Sofosbuvir Plus Daclatasvir Plus Simeprevir
12wk with ribavirin (F0–F2) or 24wk with ribavirin (F3–F4)	No	No	No	No	No
12wk with ribavirin (F0–F2) or 24wk with ribavirin (F3–F4)	No	No	No	No	No
12wk with ribavirin (F0–F2) or 24wk with ribavirin (F3–F4)	No	No	No	No	No
No	No	24wk with ribavirin	No	24wk with ribavirin	24wk with ribavirin
No	No	12wk with ribavirin (F0–F2) or 24wk with ribavirin (F3–F4)	No	12wk with ribavirin (F0–F2) or 24wk with ribavirin (F3–F4)	12wk with ribavirin (F0–F2) or 24wk with ribavirin (F3–F4)
No	No	No	No	No	No
No	No	No	No	No	No
No	No	No	12wk with ribavirin (F0–F2) or 24wk with ribavirin (Γ3–F4)	12wk with ribavirin (F0–F2) or 24wk with ribavirin (F3–F4)	12wk with ribavirin (F0–F2) or 24wk with ribavirin (F3–F4)
No	No	No	No	No	No

DRUG–DRUG INTERACTIONS BETWEEN ANTIRETROVIRALS AND DAAS

Most persons coinfected with HCV and HIV are taking multidrug ART, which may pose a problem with drug–drug interactions when initiating therapy with HCV medications. The drug–drug interactions between antiretroviral and DAAs are summarized as follows:

Daclatasvir: The NS5A inhibitor daclatasvir is a substrate of CYP3A. When daclatasvir is given with a CYP3A inhibitor, the levels of daclatasvir can increase, particularly with strong inhibitors of CYP3A. The dose of daclatasvir should therefore be reduced to 30 mg when used with either ritonavir-boosted atazanavir or lopinavir. In contrast, when used with efavirenz, a CYP3A inducer, the dose of daclatasvir should be increased to 90 mg daily.

Ledipasvir–Sofosbuvir: The NS5A inhibitor ledipasvir is not metabolized by the cytochrome p450 system but is a substrate of p-glycoprotein. Ledipasvir increases tenofovir levels by 1.3- to 2.6-fold when concomitantly given with either rilpivirine or efavirenz. Although ledipasvir administered concomitantly with tenofovir and an HIV protease inhibitor has not been studied, there is concern that tenofovir levels may increase substantially with this combination. Because of this concern and lack of data, the use of ledipasvir with ritonavir-boosted HIV protease inhibitors should, if possible, be avoided. For similar reasons, ledipasvir–sofosbuvir should not be used with cobicistat, elvitegravir, or tipranavir. Ledipasvir–sofosbuvir should not be used in HIV-infected patients on tenofovir if the baseline creatinine clearance is less than 60 mL/min.

Ombitasvir–Paritaprevir–Ritonavir: The major concern for drug interaction with this regimen is the significant p450 inhibition generated by ritonavir. This combination regimen should not be used with efavirenz, rilpivirine, darunavir, or lopinavir–ritonavir.

Ombitasvir–Paritaprevir–Ritonavir and Dasabuvir: The major concern for drug interaction with this regimen is the significant p450 inhibition generated by ritonavir. This combination regimen should not be used with efavirenz, rilpivirine, darunavir, or lopinavir–ritonavir.

Peg-interferon alfa: The metabolism of peg-interferon alfa occurs predominantly via CYP1A2. No major drug–drug interactions exist with peg-interferon and antiretroviral medications. Ribavirin: Significant and serious toxic drug–drug interactions and severe toxicities can occur with the simultaneous use of ribavirin and certain HIV NRTIs. The use of ribavirin with didanosine is strictly contraindicated because of a marked increase in intracellular didanosine levels, which may cause hepatic failure, pancreatitis, and lactic acidosis. This can also occur with stavudine or zidovudine. Thus, simultaneous use of ribavirin with didanosine, stavudine, or zidovudine should be avoided. Concurrent use of ribavirin and zidovudine should also be avoided because of additive hematologic toxicity and increased risk of severe anemia with this combination.

Simeprevir: This NS34A HCV protease inhibitor has complex interactions with antiretroviral medications because it is a substrate and an inhibitor of CYP3A4 and p-glycoprotein. In addition, simeprevir inhibits the OATP1B1/3 drug transporter. Simeprevir should not be used concomitantly with any of the following medications: efavirenz, etravirine, nevirapine, any HIV protease inhibitors, or any regimen that contains cobicistat. Simeprevir can be used with reverse transcriptase inhibitors, rilpivirine, dolutegravir, and raltegravir; if used with maraviroc, the dose of maraviroc should be decreased to 150 mg twice daily.

Sofosbuvir: This NS5B polymerase inhibitor is rapidly converted to a dominant circulating metabolite (GS-331,007). Sofosbuvir is not metabolized by the cytochrome p450 system, but is a substrate of p-glycoprotein. The only significant interaction with antiretroviral medications occurs with the p-glycoprotein inducer tipranavir, which may decrease the levels of sofosbuvir and the GS-331,007 metabolite. Accordingly, sofosbuvir should not be used concomitantly with tipranavir, but it can be use with all other antiretrovirals.

CONCLUSION

In patients with chronic hepatitis C, coinfection with HIV can accelerate the progression of hepatic fibrosis. Therefore, treatment of HCV should have high priority in coinfected patients. The introduction of DAAs, and in particular interferon-free combination therapy, has changed the landscape of therapy for patients coinfected with HCV and HIV, with multiple studies demonstrating comparable rates of sustained virologic response in coinfected and monoinfected patients. The recent HCV treatment guidelines recommend using the same HCV treatment approach for patients coinfected with HIV as that used for those with HCV monoinfection, except that in coinfected patients special consideration should be given to monitoring and managing drug–drug interactions. The use of peg-interferon and ribavirin alone or peg-interferon and ribavirin plus either boceprevir or telaprevir is no longer recommended for the treatment of HCV in coinfected patients (or HCV-monoinfected patients) or in HIV-infected patients (or HCV-monoinfected patients). ART may slow liver disease progression in HIV/HCV-coinfected patients and should therefore be considered for all coinfected patients regardless of CD4 cell count. Despite the high efficacy of DAAs, these regimens pose a huge challenge for treating coinfected patients in resource-limited countries.

ABBREVIATIONS

HCV Hepatitis C virus
HIV Human immunodeficiency virus

REFERENCES

[1] HIV data. UNAIDS. 2016. [Internet]. Available at: http://www.unaids.org/en/resources/documents/2017/AIDSdata2016.

[2] HIV data. WHO. 2016. [Internet]. Available at: http://www.who.int/hiv/en/.

[3] Global HIV and AIDS statistics | AVERT [Internet]. Available at: https://www.avert.org.

[4] Beyrer C, Baral S, van Griensven F, Goodreau SM, Chariyalertsak S, Wirtz A, Brookmeyer R. Global epidemiology of HIV infection in men who have sex with men. The Lancet July 28, 2012;380.

[5] Lowndes C, Alary M, Belleau M, Kofi Bosu W, Frédéric Kintin D, Asonye Nnorom J, et al. West Africa HIV/AIDS epidemiology and response synthesis characterisation of the HIV epidemic and response in west Africa: implications for prevention. World Bank Global HIV/AIDS Program Report. 2008.

[6] Piot P, Abdool Karim SS, Hecht R, et al. Defeating AIDS-advancing global health. Lancet 2015;386(9989):171–218.

[7] Iwuji C, Newell ML. Towards control of the global HIV epidemic: addressing the middle-90 challenge in the UNAIDS 90-90-90 target. PLoS Med May 2, 2017;14(5):e1002293.

[8] Tanser F, Bärnighausen T, Grapsa E, Zaidi J, Newell ML. High coverage of ART associated with decline in risk of HIV acquisition in rural KwaZulu-Natal, South Africa. Science 2013;339(6122):966–71.

[9] Joint United Nations Programme on HIV/AIDS (UNAIDS). The Gap Report. 2014. ISBN: 978-92-9253-062-4.

[10] Herbst AJ, Cooke GS, Bärnighausen T, KanyKany A, Tanser F, Newell ML. Adult mortality and antiretroviral treatment roll out in rural KwaZulu-Natal, South Africa. Bull World Health Organ 2009;87(10):754–62.

[11] Mossong J, Grapsa E, Tanser F, Bärnighausen T, Newell ML. Modelling HIV incidence and survival from age specific seroprevalence after antiretroviral treatment scale up in rural South Africa. AIDS 2013;27(15):2471–9.

[12] Nsanzimana S, Remera E, Kanters S, et al. Life expectancy among HIV positive patients in Rwanda: a retrospective observational cohort study. Lancet Glob Health 2015;3(3):e169–77.

[13] Bor J, Herbst AJ, Newell ML, Bärnighausen T. Increases in adult life expectancy in rural South Africa: valuing the scale up of HIV treatment. Science 2013;339(6122):961–5.

[14] Mills EJ, Bakanda C, Birungi J, et al. Mortality by baseline CD4 cell count among HIV patients initiating antiretroviral therapy: evidence from a large cohort in Uganda. AIDS 2011;25(6):851–5.

[15] Mills EJ, Bakanda C, Birungi J, et al. Life expectancy of persons receiving combination antiretroviral therapy in low-income countries: a cohort analysis from Uganda. Ann Intern Med 2011;155(4):209–16.

[16] Marston M, Michael D, Wringe A, et al. The impact of antiretroviral therapy on adult mortality in rural Tanzania. Trop Med Int Health 2012;17(8):e58–65.

[17] Reniers G, Slaymaker E, Nakiyingi-Miiro J, et al. Mortality trends in the era of antiretroviral therapy: evidence from the network for longitudinal population based HIV/AIDS data on Africa (ALPHA). AIDS 2014;28(Suppl. 4):S533–42.

[18] WHO. New hepatitis data highlight need for urgent global response [Internet].http://www.who.int/mediacentre/news/releases/2017/global-hepatitis-report/en/.

[19] Medland NA, Chow EP, Bradshaw CS, Read TH, Sasadeusz JJ, Fairley CK. Predictors and incidence of sexually transmitted Hepatitis C virus infection in HIV positive men who have sex with men. BMC Infect Dis March 2, 2017;17(1):185.

[20] Gerberding JL. Incidence and prevalence of human immunodeficiency virus, hepatitis B virus, hepatitis C virus, and cytomegalovirus among health care personnel at risk for blood exposure: final report from a longitudinal study. J Infect Dis 1994;170:1410–7.

[21] Rotman Y, Liang TJ. Coinfection with hepatitis C virus and human immunodeficiency virus: virological, immunological, and clinical outcomes. J Virol 2009;83:7366–74.

[22] Koziel MJ, Peters MG. Viral hepatitis in HIV infection. N Engl J Med 2007;356:1445–54.

[23] Allory Y, Charlotte F, Benhamou Y, Opolon P, Le Charpentier Y, Poynard T. Impact of human immunodeficiency virus infection virus infection on the histological features of chronic hepatitis C: a case-control study. The MULTIVIRC group. Hum Pathol 2000;31:69–74.

[24] Kim AY, Chung RT. Coinfection with HIV-1 and HCV—a one-two punch. Gastroenterology 2009;137:795–814.

[25] Benhamou Y, Bochet M, Di Martino V, et al. Liver fibrosis progression in human immunodeficiency virus and hepatitis C virus co-infected patients. Hepatology 1999;30:1054–8.

[26] Tefler P, Sabin C, Delvereux H, Scott F, Dusheiko G, Lee C. The progression of HCV-associated liver disease in a cohort of haemophilic patients. Br J Haematol 1994;87:555–61.

[27] Eyster ME, Diamondstone LS, Lien JM, et al. Natural history of hepatitis C virus infection in multitransfused hemophiliacs: effects of coinfection with human immunodeficiency virus. J Acquir Immune Defic Syndr 1993;6:602–10.

[28] Lesens O, Deschenes M, Steben M, Belanger G, Tsoukas C. Hepatitis C virus is related to progressive liver disease in human immunodeficiency virus-positive hemophiliacs and should be treated as an opportunistic infection. J Infect Dis 1999;179:1254–8.

[29] Soto B, Sanchez-Quijano A, Rodrigo L, et al. HIV infection modifies the nature history of chronic parenterally acquired hepatitis C with an unusually rapid progression to cirrhosis. A multicenter study on 547 patients. J Hepatol 1997;26:1–5.

[30] The French METAVIR Cooperative Study Group. Interobserver and intraobserver variation in liver biopsy interpretation in patients with chronic hepatitis C. Hepatology 1994;20:15–20.

[31] Bica I, McGovern B, Dhar R, Stone D, McGowan K, Scheib R, Snydman DR. Increasing mortality due to end-stage liver disease in patients with human immunodeficiency virus infection. Clin Infect Dis 2001;32:492–7.

[32] Arnold DM, Julian JA, Walker IR, Association of Hemophilia Clinic Directors of Canada. Mortality rates and causes of death among all HIV-positive individuals with hemophilia in Canada over 21 years of follow-up. Blood 2006;108:460–4.

[33] Williams I, Churchill D, Anderson J, Boffito M, Bower M, Cairns G, Cwynarski K, Edwards S, Fidler S, Fisher M, Freedman A, Geretti AM, Gilleece Y, Horne R, Johnson M, Khoo S, Leen C, Marshall N, Nelson M, Orkin C, Paton N, Phillips A, Post F, Pozniak A, Sabin C, Trevelion R, Ustianowski A, Walsh J, Waters L, Wilkins E, Winston A, Youle M. British HIV Association guidelines for the treatment of HIV-1-positive adults with antiretroviral therapy. 2014.

[34] Lange CG, Lederman MM. Immune reconstitution with antiretroviral therapies in chronic HIV-1 infection. J Antimicrob Chemother 2003;51:1–4.

[35] Brau N, Salvatore M, Rios-Bedoya CF, Fernandez-Carbia A, Paronetto F, Rodriguez-Orengo JF, Rodríguez-Torres M. Slower fibrosis progression in HIV/HCV-coinfected patients with successful HIV suppression using antiretroviral therapy. J Hepatol 2006;44:4755.

[36] Bonacini M. Liver injury during highly active antiretroviral therapy: the effect of hepatitis C coinfection. Clin Infect Dis 2004;38(Suppl. 2):S104–8.

[37] Sulkowski MS, Mehta SH, Torbenson M, Afdhal NH, Mirel L, Moore RD, Thomas DL. Hepatic steatosis and antiretroviral drug use among adults coinfected with HIV and hepatitis C virus. AIDS 2007;19:585–92.

[38] Vento S, Garofano T, Renzini C, Casali F, Ferraro T, Concia E. Enhancement of hepatitis C virus replication and liver damage in HIV-coinfected patients on antiretroviral combination therapy. AIDS 1998;12:116–7.

[39] Ragni MV, Bontempo FA. Increase in hepatitis C virus load in hemophiliacs during treatment with highly active antiretroviral therapy. J Infect Dis 1999;180:2027–9.

[40] Verma S, Wang CH, Govindarajan S, Kanel G, Squires K, Bonacini M. Do type and duration of antiretroviral therapy attenuate liver fibrosis in HIV-hepatitis C virus coinfected patients? Clin Infect Dis 2006;42:262–70.

[41] AASLD/IDSA. Recommendations for testing, management, and treating hepatitis C. Unique patient populations: patients with HIV/HCV confection.

[42] EASJ. Recommendations for testing, management, and treating hepatitis C. Unique patient populations: patients with HIV/HCV confection. 2016. [Internet]. Available at: http://www.easl.eu/medias/cpg/HCV2016/Summary.pdf.

[43] Kohli A, Osinusi A, Sims Z, et al. Virological response after 6 week triple-drug regimens for hepatitis C: a proof-of-concept phase 2A cohort study. Lancet 2015;385:1107–13.

[44] Kowdley KV, Gordon SC, Reddy KR, et al. Ledipasvir and sofosbuvir for 8 or 12 weeks for chronic HCV without cirrhosis. N Engl J Med 2014;370:1879–88.

[45] Laguno M, Cifuentes C, Murillas J, et al. Randomized trial comparing pegylated interferon alpha-2b versus pegylated interferon alpha-2a, both plus ribavirin, to treat chronic hepatitis C in human immunodeficiency virus patients. Hepatology 2009;49:22–31.

[46] Martel-Laferrière V, Brinkley S, Bichoupan K, et al. Virological response rates for telaprevir-based hepatitis C triple therapy in patients with and without HIV coinfection. HIV Med 2013;15:108–15.

[47] Molina JM, Orkin C, Iser DM, et al. Sofosbuvir plus ribavirin for treatment of hepatitis C virus in patients co-infected with HIV (PHOTON-2): a multicentre, open-label, non-randomised, phase 3 study. Lancet 2015;385:1098–106.

FURTHER READING

[1] Maartens G, Celum C, Lewin SR. HIV infection: epidemiology, pathogenesis, treatment, and prevention. Lancet 2014;384(9939):258–71.

[2] Naggie S, Cooper C, Saag M, et al. Ledipasvir and sofosbuvir for HCV in patients coinfected with HIV-1. N Engl J Med 2015;373:705–13.

[3] Núñez M, Miralles C, Berdún MA, et al. Role of weight-based ribavirin dosing and extended duration of therapy in chronic hepatitis C in HIV-infected patients: the PRESCO trial. AIDS Res Hum Retroviruses 2007;23:972–82.

[4] Osinusi A, Meissner EG, Lee YJ, et al. Sofosbuvir and ribavirin for hepatitis C genotype 1 in patients with unfavorable treatment characteristics: a randomized clinical trial. JAMA 2013;310:804–11.

[5] Osinusi A, Townsend K, Kohli A, et al. Virologic response following combined ledipasvir and sofosbuvir administration in patients with HCV genotype 1 and HIV co-infection. JAMA 2015;313:1232–9.

[6] Piroth L, Paniez H, Taburet AM, et al. High cure rate with 24 weeks of daclatasvir-based quadruple therapy in treatment-experienced, null-responder patients with HIV/hepatitis C virus genotype 1/4 coinfection: the ANRS HC30 QUADRIH study. Clin Infect Dis 2015.

[7] (a) Rockstroh JK, Nelson M, Katlama C, et al. Efficacy and safety of grazoprevir (MK-5172) and elbasvir (MK-8742) in patients with hepatitis C virus and HIV co-infection (C-EDGE CO-INFECTION): a non-randomised, open-label trial. Lancet HIV 2015;2:e319–27.

(b) Rockstroh JK, Spengler U. HIV and hepatitis C virus co-infection. Lancet Infect Dis 2004.

[8] Rodriguez-Torres M, Gaggar A, Shen G, et al. Sofosbuvir for chronic hepatitis C virus infection genotype 1–4 in patients coinfected with HIV. J Acquir Immune Defic Syndr 2015;68:543–9.

[9] Sulkowski M, Hezode C, Gerstoft J, et al. Efficacy and safety of 8 weeks versus 12 weeks of treatment with grazoprevir (MK-5172) and elbasvir (MK-8742) with or without ribavirin in patients with hepatitis C virus genotype 1 mono-infection and HIV/hepatitis C virus co- infection (C-WORTHY): a randomised, open-label phase 2 trial. Lancet 2015;385:1087–97.

[10] Sulkowski M, Pol S, Mallolas J, et al. Boceprevir versus placebo with pegylated interferon alfa-2b and ribavirin for treatment of hepatitis C virus genotype 1 in patients with HIV: a randomised, double-blind, controlled phase 2 trial. Lancet Infect Dis 2013;13:597–605.

[11] Sulkowski MS, Eron JJ, Wyles D, et al. Ombitasvir, paritaprevir co-dosed with ritonavir, das-abuvir, and ribavirin for hepatitis C in patients co-infected with HIV-1: a randomized trial. JAMA 2015;313:1223–31.

Chapter 4.4

Hepatitis B and C Coinfection

Georgios Zacharakis[1,2]
[1]*University of Nicosia, Nicosia, Republic of Cyprus;* [2]*Prince Sattam bin Abdulaziz University, Al Kharj, Saudi Arabia*

Chapter Outline

INTRODUCTION

Chronic hepatitis B virus (HBV) and hepatitis C virus (HCV) infections are major public health issues globally [1]. The primary concern with chronic HBV and HCV coinfection is that coinfection seems to result in more severe liver disease progression than monoinfections [2–4], with an increased risk of hepatocellular cancer (HCC) [5–7] and of fulminant hepatitis [8]. New findings confirmed that HBV/HCV coinfected patients may experience more rapid progression of severe liver disease compared with those with hepatitis B alone, although similar data for Asian and African patients with the highest prevalence

Hepatitis C in Developing Countries. https://doi.org/10.1016/B978-0-12-803233-6.00013-8

of HBV infection worldwide are still missing [9]. Furthermore, coinfection did not worsen HCV liver disease progression in the same study [9]. Nevertheless, treatment should be prioritized for HCV/HBV dually infected patients to avoid severe liver disease progression [10]. European Association for the Study of the Liver (EASL) suggested that the same therapeutic criteria should be applied to patients who are coinfected based on the viral dominance after hepatitis B and C viral interactions are applied to patients with HBV or HCV infection alone [10]. The introduction of new direct-acting antivirals (DAAs) has brought new therapeutic options in treating HCV, which need to be evaluated in coinfected patients with both HBV and HCV. But it seems reasonable as a first-line therapy based on existing data for DAA therapies in HCV monoinfected patients. A shift to a high-potency and high–genetic barrier nucleos(t)ide analogue (NA) for patients with HBV DNA persistence should always be considered. Rarely, both HBV and HCV treatment could be implemented in case of HBV reactivation during or after HCV treatment or synchronously active HBV/HCV infection [10]. The aim of this chapter is to summarize the risk factors of transmission of dual infection of hepatitis B and C and its epidemiology, to describe clinical characteristics of its liver disease with the therapeutic options for this special category of patients.

EPIDEMIOLOGY

It has been estimated that HBV carriers are 400 million people worldwide with 75% of them living in Asia and the Western Pacific [11]. About 170 million people have HCV infection globally [11–13], and there is a marked geographic variation, ranging from 1.3% to 1.6% in the United States to 15% in Egypt [14]. Both hepatotropic viruses, HBV and HCV, are transmitted parenterally, and subsequently, the coinfection rate is not uncommon in highly endemic countries in Asia, sub-Saharan Africa, and South America and among subjects with a high risk of parenteral transmission [14]. However, the true prevalence of HBV/HCV coinfection is unknown. This is because of lack of large epidemiology studies and the unknown prevalence of occult hepatitis B (OHB) infection [negative hepatitis B surface antigen (HBsAg) but detectable serum HBV DNA levels with or without serologic markers of previous viral exposure] in patients with chronic HCV infection.

Several studies reported different coinfection rates depending on the geographical region. The prevalence rates are 0.7% in Egypt [15], 16% in India [16], and 2.6% in Turkey [17]. Also, data from other countries, such as Italy [18,19], Spain [20], Japan [6,21], Taiwan [22], and Iran [23], have demonstrated that approximately 10%–15% of patients with chronic HBV infection are also infected with HCV. But, about 2%–10% of anti-HCV–positive patients are HBsAg positive.

Obviously, the risk factors of coinfection are similar to those of HBV or HCV infection alone [24–28]. Transmission of HBV through the sharing of

TABLE 4.4.1 Prevention of HBV/HCV Co-infection: WHO Recommendations

People who inject drugs: the rapid hepatitis B vaccination regimen should be offered.

People who inject drugs: incentives should complete the hepatitis B vaccination schedule.

Implement sterile needle and syringe programs for people who inject drugs.

Life style changes such as avoid sharing personal items, such as razors or toothbrushes, sharing drug needles or other drug equipment (such as straws for snorting drugs), be cautious in getting tattoos and body piercings

Offer opioid substitution therapy to treat opioid dependence; reduce HCV risk behavior and transmission through injecting drug use; and increase adherence to HCV treatment.

Integrate treatment of opioid dependence with medical services for hepatitis.

HBV, hepatitis B virus; *HCV*, hepatitis C virus.

contaminated needles among people who inject drugs (PWID) is an important route of HBV and HCV transmission in many countries of PWID with HCV infection; 25% may also have HBV infection [29–31]. Also people on hemodialysis, those who undergo organ transplantation, and people who are HIV-positive and have β-thalassemia are at high risk. Other risk factors are unsafe sex and other modes of parenteral transmission. Coinfected patients are at higher risk of developing HCC [6], especially those at a younger age [32,33]. Therefore, preventive measures are essential for care of these special populations to reduce this risk of transmission. WHO recommendations [34] reported for prevention from HBV and HCV infection among PWID are summarized in Table 4.4.1.

VIRAL INTERACTIONS

The virologic interactions of HBV/HCV coinfection are poorly understood. HCV is an RNA pathogen, whereas HBV is a DNA viral pathogen with a different life cycle. Coinfection mostly leads to the viral suppression of one of the agents. Two major specific patterns of infection have been described: the HCV superinfection of a chronic HBV-infected individual followed by suppression of the HBV replication [35] and the acute HBV infection of a patient with chronic HCV infection followed by inhibition of HCV replication [4]. This inverse relationship of the replicative patterns of the two viruses suggests direct or indirect (i.e., mediated by host immune responses) viral interference [36,37].

The viral interactions between hepatitis B and hepatitis C coinfected patients have been investigated in in vitro experiments and in clinical studies. For the in vitro experiments, the Huh-7 human HCC cell line was used to show that

HBV and HCV could replicate in vitro in the same hepatocyte without evidence of interference [38,39]. So the viral interactions observed in dually infected patients were probably because of indirect mechanisms mediated by innate and/ or adaptive host immune responses.

Regarding the suppression of HCV replication by HBV, this been shown in in vivo studies in chimpanzees and in mouse models with acute and chronic HBV infection. In chimpanzees, acute HCV superinfection in chronic HBV infection resulted in marked reduction in serum HBsAg levels [40,41]. Moreover, in mouse models, the HCV core inhibited hepatitis B replication and HBsAg expression [42].

In clinical setting also, few studies investigated the mechanisms for the inhibition of HBV replication by HCV through the HCV core protein but still are not well understood [43,44]. The HCV core protein acts as a repressor of HBV replication. Shih et al. [43] reported that HBV mRNA and HBV antigen expression had a moderate 2- to 4-fold reduction in the presence of HCV structure genes and a stronger suppression of HBV particle secretion up to 20-fold. In another study [44] this HCV core protein showed suppression of the HBV enhancer 1, up to 11-fold and the enhancer 2 up to 3- to 4-fold. Furthermore, the inhibitory effect of HCV by HCV core protein on HBV enhancer 1 was genotype-dependent being stronger in genotype 1b than genotype 3a or 1a. However, until now it not clear if or how HCV core may be released from the replication/ translation complex of HCV, but it is the first step required for many reported in vitro activities of isolated core expression systems.

In an Italian study [37], 133 untreated patients with HBV/HCV coinfection were followed up for 1 year with measurement of serum HBV DNA and HCV RNA levels and liver biochemistry. Approximately one-third of the patients appeared with phases of suppression and reactivation of the replication activity of one or both viruses suggesting that HBV/HCV coinfection has a dynamic viral profile.

Furthermore, patients with chronic HBV infection superinfected with HCV may appear to have hepatitis B envelope antigen (HBeAg) and HBsAg seroconversion [45–47]. Sheen et al. [48] estimated an annual rate of 2.08% of HBsAg seroconversion in HBV/HCV coinfected patients compared with 0.43% in patients with HBV infection alone.

The reverse is also true: there is an inhibitory effect of HBV on HCV [5,49]. Zarski et al. [5] compared virologic characteristics of coinfected HBV/ HCV patients with those infected with HCV alone. They showed an inverse relationship between the replicative patterns of both viruses. The HCV RNA level was significantly decreased in HBV/HCV coinfected patients with serum HBV DNA–positive patients as compared with HBV DNA–negative patients. Another study showed that coinfected patients had a higher rate of HCV RNA clearance compared with those with HCV infection alone (71% vs. 14%) [50].

In real life, the alternating replicative patterns of both viruses in coinfected HBV/HCV patients have a clinical impact. This is often because one of the viruses is replicating at a much faster rate and inhibiting the replication of the other. Indeed, in some patients with HBV/HCV dual infection after eradication of the dominant virus such as clearance of serum HCV RNA with pegylated interferon-α (Peg-IFN-α) and ribavirin (RBV), the other virus then may become active (HBV reactivation) [51,52]. An interesting case of HBV reactivation with the DAAs for HCV therapy was reported in the study of Takayama et al. [53]. A dually infected patient received initial treatment with asunaprevir and daclatasvir for the HCV dominance, resulting in HCV early virologic response but an increase in HBV DNA, requiring the use of a NA entecavir as shown in Fig. 4.4.1.

In summary, the majority of HBV/HCV coinfected patients may have an active HCV and inactive HBV replication phase. Some other patients experience high HBV DNA levels but HCV RNA is undetectable, while others present with alternating phases of dominance of one virus over the other [54].

Clinical Features of Hepatitis C and Hepatitis B Coinfection

Because HBV and HCV have a common route of transmission, blood-to-blood contact, it is possible to contract both viruses at the same time. However, a person infected with one of the viruses may have a concurrent infection with the other virus later in life through the same or another mode of transmission. In either case, the problem is that coinfection can result in severe liver disease progression such as cirrhosis and/or hepatic failure and increases the risk of HCC [13].

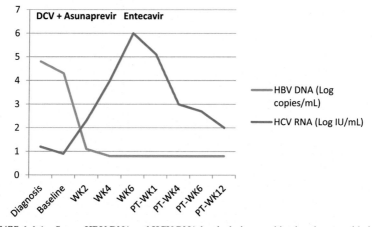

FIGURE 4.4.1 Serum HBV DNA and HCV RNA levels during combination therapy with daclatasvir and asunaprevir and 12 weeks post-treatment in an HBV/HCV coinfected patient. At baseline the patient was positive for HBsAg and anti-HCV. Serum HCV RNA levels were 4.2 log IU/mL and low HBV DNA levels were 2.5 log copies/mL. After treatment with combined regimens, serum HCV RNA levels declined. However, when HCV RNA became undetectable, HBV DNA began to rise and entecavir was added. *HBsAg*, hepatitis B surface antigen; *HBV*, hepatitis B virus; *HCV*, hepatitis C virus.

Pol et al. [9] studied the characteristics and clinical outcomes among patients with dual infection of HBV/HCV in the ANRS CO22 HEPATHER cohort, comparing them with participants with either hepatitis B or C monoinfection. The ANRS CO22 HEPATHER cohort is a large cohort of people with hepatitis B and/or C followed up by more than 30 centers in France. The authors studied 14,698 participants. Of these participants, 1,099 proved to have concurrent HBV/HCV infection by serology, while 9,098 had HCV monoinfection and 4501 had HBV monoinfection. The authors concluded that HBV/HCV coinfected patients developed a more rapid severe liver disease progression than those with HBV monoinfection but did not appear to worsen hepatitis C liver disease progression. Indeed, coinfected patients and those with HCV monoinfection had similar rates of advanced fibrosis or cirrhosis (Metavir stage F3–F4), but both were significantly higher than the rate for patients with HBV monoinfection (42% and 40% vs. 15%, respectively). This suggests that HCV treatment initiation must be prioritized in patients with an HBV/HCV coinfection, regardless of the severity of liver disease.

Regarding virologic data, coinfected patients were significantly more likely to have low serum HBV DNA levels (<50 IU/mL) than those with HBV monoinfection (87% vs. 62%), confirming that HCV inhibits HBV replication.

Another recent study by Kruse et al. [55] in 2014, a retrospective cohort study of nearly 103,000 patients with chronic HCV infection, registered at the U.S. Veterans Administration health system between 1997 and 2005 reported that 1,431 (about 1.4%) had HBV coinfection. HBV dominance was 45% of the HBV/HCV coinfected patients tested for HBV DNA. The presence of active HBV replication was associated with significant worse outcomes in coinfected patients. In contrast, absence of HBV replication was associated with clinical outcomes similar to that of HCV monoinfected patients. Furthermore, the risk of cirrhosis, HCC, or death between coinfected patients with HCV dominance and those with HCV infection alone was not significant different.

Overall more than 90% of adults with acute HBV infection develop icteric hepatitis as a result of an appropriate immune response followed by a spontaneous resolve of the virus and develop antibodies against HBV [anti–hepatitis B surface (anti-HBs)] [56,57]. Few patients develop chronic HBV hepatitis and are surface antigen (HBsAg) positive. In contrast, most adults exposed to HCV develop an acute hepatitis without icterus and approximately 80%–85% develop chronic infection [58]. These patients with a history of HCV exposure, whether they develop chronic infection or resolve HCV virus, are anti-HCV positive.

In coinfection with HBV and HCV, the following clinical outcomes may occur as shown in Fig. 4.4.2 [13]. Patients may resolve HBV and develop chronic HCV hepatitis; they may resolve hepatitis C and develop chronic HBV hepatitis; they may resolve both viruses; or they may develop chronic infection with both HBV and HCV. Clinical, serologic, and virologic characteristics of these infectious scenarios are outlined in Table 4.4.2.

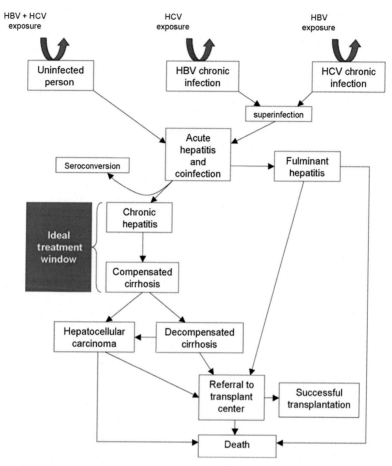

FIGURE 4.4.2 Clinical course of hepatitis B virus/hepatitis C virus coinfection.

HBV Superinfection of Patients With Chronic HCV Infection

Two entities of HBV superinfection in a patient with chronic HCV infection are seen in clinical practice: OHB infection and active HBV infection in patients with chronic HCV hepatitis. The clinical course of hepatitis C is believed to be worsened by HBV coinfection. On the other hand, it has been suggested that active HBV infection suppresses the replication of HCV.

OHB Infection in Chronic HCV Hepatitis

OHB infection is described by the presence of serum HBV DNA without detectable HBsAg with or without anti-HBc or anti-HBs outside the pre-seroconversion window period [59]. This silent HBV infection seems to occur up to 50% in patients with chronic hepatitis C without clinical significance [59–61].

TABLE 4.4.2 Clinical, Serologic, and Virologic Characteristics of Dual Infection of HBV and HCV in Patients With Chronic Hepatitis

The Clinical Spectrum	Serologic and Virologic Characteristics				
	HBsAg	HBeAg/ anti-HBe	HBV-DNA (IU/mL)	Anti-HCV	HCV RNA
Chronic HBeAg (−) hepatitis and HCV hepatitis	+	−/+	>2000	+	+
Chronic HBeAg (+) hepatitis and HCV hepatitis		+/−	>2000		
Occult HBV in chronic HCV hepatitis	−	−/−	<2000	+	+
Chronic HCV hepatitis in HBsAg carrier	+	−/+	<2000	+	+
Both viruses resolved	−	−/+	−	+	−

HBeAg, hepatitis B envelope antigen; *HBsAg*, hepatitis B surface antigen; *HBV*, hepatitis B virus; *HCV*, hepatitis C virus.

Georgiadou et al. [62] reported the presence of OHB infection (detectable serum HBV DNA of HCV-infected patients in the absence of HBsAg) in 26.2% of 540 Greek HCV patients.

Although HBV coinfection in patients with chronic HCV hepatitis can be silent, a more recent study has shown that these patients may show deterioration of chronic liver disease and may develop hepatocellular carcinoma [63]. Cacciola et al. showed that 21 of the 66 (33%) anti-HBc–positive patients with chronic HCV hepatitis had cirrhosis compared with 26 of the 134 (19.8%, $P=.04$) anti-HBc–negative patients with HCV infection alone. This suggests that integrated HBV DNA may be important to the development of hepatocellular carcinoma in HCV-positive/HBsAg-negative patients with the anti-HBc antibody [64].

Moreover, a poor response of interferon (IFN) has been described in patients with OHB coinfection and chronic HCV hepatitis [65,66]. This may be partly because of downregulation of IFN receptor gene expression in the liver. However, another study showed that virologic responses to combined Peg-IFN and RBV therapy are similar in chronic HCV patients with and without OHB infection [67].

HBV Superinfection of Patients With Chronic HCV Hepatitis

HBV superinfection was less frequently reported than HCV superinfection. It has been reported that hepatitis B can also reduce the reproduction of the HCV, although the overall dominant effect appears to be hepatitis C over the HBV [13]. Usually, patients with chronic HCV infection who have evidence of active HBV infection (HBsAg-positive, HBeAg-positive) have a significant lower prevalence of detectable serum HCV RNA (41%) compared with those who had already recovered from HBV infection (82%) [HBsAg-negative/anti–hepatitis B core (HBc)-positive] [68].

Furthermore, the histologic findings of these patients with chronic HCV infection and active HBV have shown a frequent severe liver disease than those who have HCV infection alone. There is also a higher risk for the development of HCC of these patients [64].

Finally, Sagnelli et al. [4], in a more recent, long-term follow-up study have shown that acute hepatitis B infection is more severe in 29 patients with chronic HCV and acute hepatitis B than 29 patients with acute hepatitis B infection alone (34.5% vs. 6.9%, $P < .05$). However, some patients experienced HCV RNA clearance.

Acute Coinfection of HBV/HCV

Acute coinfection of HBV and HCV is rare but more frequent in PWID [68–70]. Usually, patients showed delayed HBsAg appearance and a shorter HBs antigenemia compared with those with acute HBV infection alone [71]. Alberti [36] studied 30 patients with acute HBV/HCV hepatitis and markers of active HBV and HCV infection. All patients were followed for a long time after acute hepatitis. The chronicity rates were similar to those patients with HBV or HCV monoinfection. In acute phase, the peak of transaminases was high in patients with HBV/HCV coinfection and a biphasic alanine aminotransferase elevation was observed in some patients independent of whether hepatitis was resolved or progressed to chronicity. Furthermore, spontaneous clearance of either or both viruses has been documented in the literature [36,72–74].

HCV Superinfection of Patients With Chronic HBV Infection

The phenomenon of HCV superinfection is frequently described in endemic areas of HBV infection, such as Asian countries, South America, and sub-Saharan Africa [75,76]. The patients usually acquired HBV infection vertically and acquired HCV infection subsequently. So the onset of HCV infection occurs after HBV infection has been acquired. Several reports have documented that HCV superinfection can result in HBeAg seroconversion and, in some cases, reduce HBsAg expression (HBsAg loss) and may promote clearance of HBsAg [75]. The anti-HBs/anti-HBc–positive patients with HCV infection who may already have had preexisting chronic liver disease can progress to a more severe liver disease and an increased risk of fulminant hepatitis [77]. Fulminant hepatic failure was significantly higher among patients with underlying HBV infection than in those without (23% vs. 3%) [77,78]. Furthermore, Sagnelli et al. [4] showed that patients with HCV superinfection had a significantly higher cumulative incidence of cirrhosis and HCC than acute hepatitis Delta superinfection or active chronic HBV infection for a follow-up period of 1–21 years. The mortality rate of HCV superinfection in chronic patients infected with hepatitis B may be as high as 10% [51].

Evolution of Treatment of HBV/HCV Coinfected Patients

In patients with chronic HBV hepatitis, HCV coinfection accelerates liver disease progression and increases the risk of HCC [6,79,80]. In particular, both HBV and HCV viruses replicate in the same hepatocyte without interference [39]. Subsequently, follow up of serum HBV DNA and HCV RNA levels is necessary to determine the dominant virus that should be eradicated. It is uncommon to have a co-dominance of both viruses. HBV dominance means high serum HBV DNA levels and low serum HCV RNA levels. However, some of these patients have fluctuating serum HBV DNA levels there is a need for a long term follow-up [81] to evaluate viral loads before starting any antiviral therapy. In HCV dominance, serum HBV DNA levels are often low or undetectable and serum HCV RNA levels are high. HCV is responsible for the activity of chronic hepatitis in most coinfected patients. These patients should receive treatment for chronic HCV hepatitis [51]. However, NA treatment should be considered in patients with HBV dominance which means in patients who have active HBV replication with persistent or fluctuating serum HBV DNA levels above 2000 IU/mL. Viral interactions will affect the therapeutic options in dually infected patients with HBV and HCV depending probably on indirect mechanisms mediated by innate and/or adaptive host immune responses [39].

Treatment of HCV in Dually Infected HCV/HBV Patients With Chronic HCV Hepatitis

Standard Interferon With RBV for HCV/HBV Coinfection

Initially, interferon containing regimen has been the only therapy to eradicate HBV or HCV infection for years. Later, guanosine analog (ribavirin) was added. Indeed, in coinfected patients with HCV dominant disease, treatment with IFN plus RBV has been well studied but has proven to have poor efficacy in several studies.

In the study by Villa et al. [72], patients with HCV/HBV coinfection were treated with 9 million IU of standard IFN three times weekly for 3 months and 31% of these patients showed clearance of HCV. In their study, Liu et al. [82] treated coinfected patients with standard IFN and RBV and reported sustained virologic response (SVR) of up to 21% in those patients who lost HBsAg. SVR rates were comparable with those in patients with HCV alone.

Standard Peg-IFN With RBV for HCV/HBV Coinfection

Later, pegylation of interferon modified the pharmacokinetic profile of IFN-α-2. Compared to conventional IFN, pegylated interferon alpha showed a slower absorption rate, more reduced distribution and slower clearance rate. As a result, this molecule proven to be superior to standard interferon alpha for both chronic hepatitis B and C treatment. Treatment containing regimens with Peg-IFN plus RBV has also been well studied and has improved efficacy in several studies.

In a small, prospective, multicenter pilot study by Potthoff et al. [51], 19 patients with chronic HBV/HCV coinfection (6 were HBV DNA positive and 13 were HBV DNA negative) treated with weight-adjusted Peg-IFN-α-2β and RBV for 48 weeks achieved an SVR rate of 93% (86% in genotype 1 and 100% in genotypes 2 or 3).

Moreover, a large prospective randomized, controlled, multicenter study by Liu et al. [83] evaluated Peg-IFN and RBV in HCV/HBV coinfected patients. High SVR rates (72% and 83% in genotypes 1 and 2/3, respectively) were achieved. Also, the same authors investigated the sustained HCV clearance both in HCV-monoinfected and HCV/HBV dually infected patients by a follow-up period of 5-year [84]. They reported that HCV reappearance occurred only in 6 (2.6%) of the 232 patients who achieved SVR for a median follow-up of 4.6 ± 1.0 years and concluded that sustained HCV clearance is not influenced by HBV coinfections.

Although the above authors reported on a high effectiveness of Peg-IFN-α plus RBV in HBV/HCV-coinfections, SVR rates are affected by the HCV genotypes and the IL28B genotype.

Furthermore, adverse events are common with interferon-based regimens such as fatigue, flu-like symptoms, depression, skin rash, and gastrointestinal manifestations. Also, pegylated interferon alpha is contraindicated in patients with untreated or severe depression and with decompensated cirrhosis. These factors have driven the urgent need for more effective and safer such as the DAAs including inhibitors of nonstructural proteins such as NS3/4A protease, NS5A and NS5B polymerase for the treatment of chronic hepatitis C.

DAAs greatly improve SVR rates independently of whether the patient is treatment naïve or previously unsuccessfully treated with peginterferon and ribavirin, of fibrosis stage and IL28B genotype. Subsequently, therapeutic management of HCV-infected adults includes only DAAs [10].

New therapies including regimens with high potency and high genetic barrier were administered orally and this resulted in better compliance and high rates of virological responses for eradication of HCV.

DAAs for HCV/HBV Co-infection

At the moment, little data are available whether the new DAA-based therapies will be effective in HBV/HCV coinfection. Collins et al. [85] reported two coinfected HCV/HBV cases treated with potent, IFN-free, DAA regimens simeprevir and sofosbuvir with no activity against HBV who had HBV reactivation. They suggested that early initiation of anti-HBV therapy in the setting of increasing serum HBV DNA levels should be strongly considered to prevent significant hepatitis as recommended by EASL guidelines [10]. Current HCV treatment guidelines are now clear about how to monitor these patients with an unpredictable risk of HBV reactivation. Serum aminotransferase levels monitoring is indicated in patients with resolved HBV infection (anti-HBs and anti-HBc antibody-positive patients). If serum ALT levels elevated on treatment a test for HBs antigen and/or serum HBV DNA levels should be done. In case of chronic hepatitis B or "occult" HBV infection the addition of NA therapy is strongly recommended. Concurrent HBV therapy with DAAs combined treatment for HCV is strongly recommended if serum HBV DNA levels are detectable at a significant level before or after initiation of HCV treatment, or if there is a potential risk of HBV reactivation such as after initiation of a wide variety of immune-suppressive therapies for the treatment of cancer or autoimmune disease and in patients who are solid-organ transplant recipients [10]. Previously all HCV treatment regimens included Peg-IFN-α, which also has significant activity against HBV [10,86,87]. The risk of HBV reactivation may be greater with the newer DAAs, given their high potency against HCV and lack of anti-HBV activity. Few interesting cases of HBV reactivation with the DAAs for HCV therapy were recently reported [87a]. For the period from November 2013 to October 2016, the US Food and Drug Administration (FDA) has reported 29 cases of HBV reactivation in patients with dual HCV/HBV infection who received DAAs regimens as anti-HCV therapy [87b]. As a result, FDA now issued a warning about the risk of HBV reactivation in those patients who are going to be treated with DAAs. Subsequently, routine initiation of dual treatment for HCV and HBV might be the standard of care for all coinfected patients [86].

An ongoing phase 3 trial (NCT02555943) [88] is currently assessing the incidence, morbidity, mortality, and predisposing factors for the reactivation of HBV replication during direct anti-HCV treatment of HCV/HBV coinfection patients. In one arm HCV/HBV coinfection patients will receive NA (Entecavir or Tenofovir disoproxil fumarate) for the treatment of hepatitis B infection when HBV viral breakthrough occurred during anti-HCV treatment using DAAs (Ledipasvir/Sofosbuvir; or Sofosbuvir and Daclatasvir, or Ombitasvir, Paritaprevir, Ritonavir, Dasabuvir; or Sofosbuvir + Ribavirin). In the other arm, HCV/HBV coinfection patients will receive NA (Entecavir or Tenofovir disoproxil fumarate) for the treatment of hepatitis B infection before or at the commencement of direct anti-HCV treatment using DAAs (Ledipasvir/Sofosbuvir; or Sofosbuvir and Daclatasvir, or Ombitasvir, Paritaprevir, Ritonavir, Dasabuvir; or Sofosbuvir + Ribavirin).

Moreover, IFN-based HCV treatment can result in loss of HBV suppression and HBV reactivation. A recent meta-analysis showed that dually infected patients treated with Peg-IFN-α and RBV appeared with an increase in HBV replication (23%) and particularly those with SVR to anti-HCV therapy [89]. But, this phenomenon is not frequent [9], and acute hepatitis appears to be more rare [90].

There are still unresolved issues and unmet needs regarding treatment of dually infected patients with DAAs in combination with DAAs, Peg-IFN plus RBV or with IFN-free regimens. There is a need to develop effective and optimum treatment for HBV/HCV coinfection.

Treatment of HBV in Dually Infected HCV/HBV Patients With Chronic HBV Hepatitis

Peg-IFN–based therapy in HCV/HBV coinfected patients also acts on the HBV as does in monoinfected HBV patients [91]. Indeed, studies reported HBsAg loss 6 months after the end of therapy in 18 (11.2%) of the 161 dually infected patients with a rate of HBsAg seroclearance of 5.4% per year [83,84]. Baseline low serum HBsAg levels correlated significantly with HBsAg seroclearance. Another study [92] showed that the host genetic polymorphism rs9277535 for HLA-DPB1 region was associated with spontaneous HBsAg seroclearance.

However, during treatment or after clearance of HCV, reactivation of HBV because of immune reconstitution may appear [51]. In a study, in patients treated with baseline serum HBV DNA levels <200 IU/mL, HBV DNA reappearance occurred in 47 (61.8%) of 76 patients [84]. Therefore, serum HBV DNA level monitoring is necessary and in case of HBV, reactivation treatment with NA(s) is necessary.

Little data are available on the use of anti-HBV drugs for patients coinfected with HBV/HCV and hepatitis B–dominant disease.

IFN with or without Lamivudine (LAM) was a reasonable option in the past, before the new antiviral molecules become available, for coinfected patients. In a small study [93], eight patients with dually active HBV and HCV were treated with 5 MU IFN and 100 mg/day LAM for 12 months, followed by LAM alone for 6 more months. The SVR for HCV was 50%. HBeAg clearance was observed in three patients, and two of them seroconverted to anti-HBe. Serum HBV DNA levels were undetectable in three patients at the end of treatment, but appeared again later in two patients.

Today coinfected patients should receive a potent antiviral agent with a high barrier to resistance (i.e., entecavir or tenofovir) to reduce levels of HBV DNA ideally to undetectable or at least to <2000 IU/mL as recommended by EASL guidelines [91].

In a more recent study, Coppola et al. [94] studied in a cohort of 24 HBV/HCV coinfected cirrhotic patients the tolerability and efficacy of anti-HBV NAs [lamivudine plus adefovir (n = 10), entecavir (n = 7), telbivudine (n = 4), and tenofovir disoproxil fumarate (n = 3)]. Undetectable serum HBV DNA levels were found in 96% of patients after 18 months, while HCV reactivation was low (12.5%). But, 33% of these patients appeared with progression of liver cirrhosis.

Serum HCV RNA–positive patients at baseline deteriorated more frequently. The authors concluded that a favorable clinical outcome of HBV/HCV cirrhotic patients was seen only in HCV RNA–negative patients at baseline. Today, after pegylated interferon discontinuation treatment of chronic HBV/HCV with HBV dominance should subsequently be with NAs. Over the last several years, several new agents have been added to the armamentarium of drugs against HBV infection. Currently, a potent antiviral NA with high genetic barrier i.e., entecavir, tenofovir disoproxil, or tenofovir alafenamide, represents the optimal agents for first-line therapy [91]. Other NAs such as lamivudine, adefovir, and telbivudine are not recommended due to the high viral resistance.

Guidelines for the Treatment of Coinfected HBV/HCV Patients

Effective treatment should eradicate HCV infection and inhibit HBV replication. Evaluation of liver disease progression, predominance of one virus over another, and comorbidities are essential for optimal antiviral regimens. For patients with active hepatitis C, the same regimens following the same rules as for monoinfected patients should be applied based on EASL recommendations [10]. Regimens such as Peg-INF plus RBV with or without a second-generation DAA or DAA-based and IFN-free regimens should be implemented [10,95]. Hopefully, the high cost of these drugs will not be a limitation to their use in developing countries. For HBV/HCV coinfected patients, either cirrhotics or not, with active replication of hepatitis B virus which may occur before, during or after anti-HCV treatment, NA with entecavir, tenofovir disoproxil or tenofovir alafenamide is indicated [91,96]. Concurrent HBV nucleoside/NA therapy is indicated either if there is a potential risk of HBV reactivation during or after HCV clearance or if HBV replication is detectable at a significant level before initiation of HCV treatment [10,97].

In concurrent HBV/HCV treatment with the combination of simeprevir and tenofovir, estimated glomerular filtration rate (eGFR) and tubular function should be monitored frequently during treatment because simeprevir increases exposure to tenofovir [10]. Thus, tenofovir doses should be consequently adjusted when required.

Further studies are needed to clarify if DAA-based triple therapy or all oral IFN-free regimens with or without NA are the optimal regimens of both viruses at the same time for this population.

REFERENCES

[1] El S, Pisaturo M, Martini S, Sagnelli C, Filippini P, Coppola N. Advances in the treatment of hepatitis B virus/hepatitis C virus coinfection. Expert Opin Pharmacother July 2014;15(10):1337–49.

[2] Sato S, Fujiyama S, Tanaka M, et al. Coinfection of Hepatitis C virus in patients with chronic Hepatitis B infection. J Hepatol 1994;21:159–66.

[3] Lee LP, Dai CY, Chuang WL, et al. Comparison of liver histopathology between chronic hepatitis C patients and chronic hepatitis B and C coinfected patients. J Gastroenterol Hepatol 2007;22:515–7.

[4] Sagnelli E, Coppola N, Pisaturo M, et al. HBV superinfection in HCV chronic carriers: a disease that is frequently severe but associated with the eradication of HCV. Hepatology 2009;49:1090–7.

[5] Zarski J-P, Bohn B, Bastie A, et al. Characteristics of patients with dual infection by Hepatitis B and C viruses. J Hepatol 1998;28:27–33.

[6] Donato F, Boffetta P, Puoti M. A meta-analysis of epidemiological studies on the combined effect of hepatitis B and C virus infections causing hepatocellular carcinoma. Int J Cancer 1998;75:347–54.

[7] Shi J, Zhu L, Liu S, Xie WF. A meta-analysis of case-control studies on the combined effect of hepatitis B and C virus infections in causing hepatocellular carcinoma in China. Br J Cancer 2005;92:607–12.

[8] Niederau C, Lange S, Heintges T, et al. Prognosis of chronic Hepatitis C: results of a large prospective cohort study. Hepatology 1998;28:1687–95.

[9] Pol S, Lucier S, Fontaine H, et al. Negative impact of HBV/HCV coinfection on HBV or HCV monoinfection: data from the French cohort ANRS CO22 HEPATHER. In: International liver congress: 50th annual meeting of the European Association for the Study of the Liver (EASL). Vienna, April 22–26, 2015. 2015. Abstract P0468.

[10] European Association for the Study of the Liver. EASL recommendations on treatment of hepatitis C 2016. J Hepatol 2017;66:153–94.

[11] Williams R. Global challenges in liver disease. Hepatology 2006;44:521–6.

[12] Bini EJ, Perumalswami PV. Hepatitis B virus infection among American patients with chronic hepatitis C virus infection: prevalence, racial/ethnic differences, and viral interactions. Hepatology 2010;51:759–66.

[13] Crockett SD, Keeffe EB. Natural history and treatment of hepatitis B virus and hepatitis C virus coinfection. Ann Clin Microbiol Antimicrob 2005;4:13.

[14] Karoney MJ, Siika AM. Hepatitis C Virus (HCV) infection in Africa: a review. Pan Afr Med J 2013;14:44.

[15] Ma M, Nasr AM, Saleh MA, et al. Virologic and histologic characterisation of dual hepatitis B and C co-infection in Egyptian patients. Arab J Gastroenterol 2013;14:143–7.

[16] Chakravarti A, Verma V, Jain M, Kar P. Characteristics of dual infection of hepatitis B and C viruses among patients with chronic liver disease: a study from tertiary care hospital. Trop Gastroenterol 2005;26:183–7.

[17] Senturk H, Tahan V, Canbakan B, et al. Clinicopathologic features of dual chronic hepatitis B and C infection: a comparison with single hepatitis B, C and delta infections. Ann Hepatol 2008;7:52–8.

[18] Di Marco V, Lo Iacono O, Camma C, et al. The long term course of chronic hepatitis B. Hepatology 1999;30:257–64.

[19] Fattovich G, Tagger A, Brollo L, et al. Hepatitis C virus infection in chronic hepatitis B virus carriers. J Infect Dis 1991;163:400–2.

[20] Crespo J, Lozano JL, de la CF, et al. Prevalence and significance of hepatitis C viremia in chronic active hepatitis B. Am J Gastroenterol 1994;89:1147–51.

[21] Ohkawa K, Hayashi N, Yuki N, et al. Hepatitis C virus antibody and hepatitis C virus replication in chronic hepatitis B patients. J Hepatol 1994;21:509–14.

[22] Dai CY, Yu ML, Chuang WL, et al. Influence of hepatitis C virus on the profiles of patients with chronic hepatitis B virus infection. J Gastroenterol Hepatol 2001;16:636–40.

[23] Semnani S, Roshandel G, Abdolahi N, et al. Hepatitis B/C virus co-infection in Iran: a sero-epidemiological study. Turk J Gastroenterol 2007;18:20–1.

[24] Pallas JR, Farinas-Alvarez C, Prieto D, Delgado-Rodriguez M. Coinfections by HIV, hepatitis B and hepatitis C in imprisoned injecting drug users. Eur J Epidemiol 1999;15:699–704.

[25] Reddy GA, Dakshinamurthy KV, Neelaprasad P, Gangadhar T, Lakshmi V. Prevalence of HBV and HCV dual infection in patients on haemodialysis. Indian J Med Microbiol 2005;23:41–3.

[26] Aroldi A, Lampertico P, Montagnino G, et al. Natural history of hepatitis B and C in renal allograft recipients. Transplantation 2005;79:1132–6.

[27] Kalinowska-Nowak A, Bociąga-Jasik M, Garlicki A, Skwara P. Prevalence of hepatotropic viruses HBV and HCV in HIV-infected patients from Southern region of Poland. Acta Virol 2000;44:23–8.

[28] Irshad M, Peter S. Spectrum of viral hepatitis in thalassemic children receiving multiple blood transfusions. Ind J Gastroenterol 2002;21:183–4.

[29] Pontisso P, Ruvoletto MG, Fattovich G, Chemello L, Gallorini A, Ruol A, et al. Clinical and virological profiles in patients with multiple hepatitis virus infections. Gastroenterology 1993;105(5):1529–33.

[30] Liu CJ, Liou JM, Chen DS, Chen PJ. Natural course and treatment of dual hepatitis B virus and hepatitis C virus infections. J Formos Med Assoc 2005;104(11):783–91.

[31] Potthoff A, Manns MP, Wedemeyer H. Treatment of HBV/HCV coinfection. Expert Opin Pharmacother 2010;11(6):919–28.

[32] Benvegnu L, Noventa F, Bernardinello E, Pontisso P, Gatta A, Alberti A. Evidence for an association between the aetiology of cirrhosis and pattern of hepatocellular carcinoma development. Gut 2001;48(1):110–5.

[33] Kew MC, Yu MC, Kedda MA, Coppin A, Sarkin A, Hodkinson J. The relative roles of hepatitis B and C viruses in the etiology of hepatocellular carcinoma in southern African blacks. Gastroenterology 1997;112(1):184–7.

[34] Guidance on prevention of viral hepatitis B and C among people who inject drugs. Geneva: World Health Organization; 2012. http://apps.who.int/iris/bitstream/10665/75357/1/9789241 504041_eng.pdf.

[35] Liaw YF, Chen YC, Sheen IS, Chien RN, Yeh CT, Chu CM. Impact of acute hepatitis C virus superinfection in patients with chronic hepatitis B virus infection. Gastroenterology 2004;126:1024–9.

[36] Alberti A, Pontisso P, Chemello L, et al. The interaction between hepatitis B virus and hepatitis C virus in acute and chronic liver disease. J Hepatol 1995;22:38–41.

[37] Raimondo G, Saitta C. Treatment of the hepatitis B virus and hepatitis C virus co-infection: still a challenge for the hepatologist. J Hepatol 2008;49:677–9.

[38] Eyre NS, Phillips RJ, Bowden S, et al. Hepatitis B virus and hepatitis C virus interaction in Huh-7 cells. J Hepatol 2009;51:446–57.

[39] Bellecave P, Gouttenoire J, Gajer M, et al. Hepatitis B and C virus coinfection: a novel model system reveals the absence of direct viral interference. Hepatology 2009;50:46–55.

[40] Brotman B, Prince AM, Huima T, Richardson L, van den Ende MC, Pfeifer U. Interference between non-A, non-B and hepatitis B virus infection in chimpanzees. J Med Virol 1983;11:191–205.

[41] Bradley DW, Maynard JE, McCaustland KA, Murphy BL, Cook EH, Ebert JW. Non-A, non-B hepatitis in chimpanzees: interference with acute hepatitis A virus and chronic hepatitis B virus infections. J Med Virol 1983;11:207–13.

[42] Zhu W, Wu C, Deng W, et al. Inhibition of the HCV core protein on the immune response to HBV surface antigen and on HBV gene expression and replication in vivo. PLoS One 2012;7:e45146.

[43] Shih CM, Chen CM, Chen SY, Lee YH. Modulation of the transsuppression activity of hepatitis C virus core protein by phosphorylation. J Virol 1995;69:1160–71.

[44] Schuttler CG, Fiedler N, Schmidt K, Repp R, Gerlich WH, Schaefer S. Suppression of hepatitis B virus enhancer 1 and 2 by hepatitis C virus core protein. J Hepatol 2002;37:855–62.

[45] Chu CM, Yeh CT, Liaw YF. Low-level viremia and intracellular expression of hepatitis B surface antigen (HBsAg) in HBsAg carriers with concurrent hepatitis C virus infection. J Clin Microbiol 1998;36:2084–6.

[46] Liaw YF, Tsai SL, Chang JJ, et al. Displacement of hepatitis B virus by hepatitis C virus as the cause of continuing chronic hepatitis. Gastroenterology 1994;106:1048–53.

[47] Liaw YF, Lin SM, Sheen IS, Chu CM. Acute hepatitis C virus superinfection followed by spontaneous HBeAg seroconversion and HBsAg elimination. Infection 1991;19:250–1.

[48] Sheen IS, Liaw YF, Lin DY, Chu CM. Role of hepatitis C and delta viruses in the termination of chronic hepatitis B surface antigen carrier state: a multivariate analysis in a longitudinal follow-up study. J Infect Dis 1994;170:358–61.

[49] Chen SY, Kao CF, Chen CM, et al. Mechanisms for inhibition of hepatitis B virus gene expression and replication by hepatitis C virus core protein. J Biol Chem 2003;278:591–607.

[50] Ohkawa K, Hayashi N, Yuki N, et al. Long-term follow-up of hepatitis B virus and hepatitis C virus replicative levels in chronic hepatitis patients co-infected with both viruses. J Med Virol 1995;46:258–64.

[51] Potthoff A, Wedemeyer H, Boecher WO, et al. The HEP-NET B/C co-infection trial: a prospective multicenter study to investigate the efficacy of pegylated interferon-alpha2b and ribavirin in patients with HBV/HCV co-infection. J Hepatol 2008;49:688–94.

[52] Hamzaoui L, El Bouchtili S, Siai K, Mahmoudi M, Azzouz MM. Hepatitis B virus and hepatitis C virus co-infection: a therapeutic challenge. Clin Res Hepatol Gastroenterol 2013;37:e16–20.

[53] Takayama H, Sato T, Ikeda F, Fujiki S. Reactivation of hepatitis B virus during interferon-free therapy with daclatasvir and asunaprevir in patient with hepatitis B virus/hepatitis C virus co-infection. Hepatol Res 2016;5:489–91.

[54] Pontisso P, Gerotto M, Ruvoletto MG, et al. Hepatitis C genotypes in patients with dual hepatitis B and C virus infection. J Med Virol 1996;48:157–60.

[55] Kruse R, Kramer JR, Duan Z, et al. Impact of active hepatitis B virus DNA replication on clinical outcomes in patients with hepatitis C and B Co-infection. Dig Dis Week – DDW May 3–6, 2014. Chicago. Abstract 704.

[56] McMahon BJ. The natural history of chronic hepatitis B virus infection. Hepatology May 2009;49(5 Suppl.):S45–55.

[57] Ganem D, Prince AM. Hepatitis B virus infection — natural history and clinical consequences. N Engl J Med 2004;350:1118–29.

[58] Wang LS, D'Souza LS, Jacobson IM. Hepatitis C – a clinical review. J Med Virol April 20, 2016. http://dx.doi.org/10.1002/jmv.24554.

[59] Brechot C, Thiers V, Kremsdorf D, Nalpas B, Pol S, Paterlini-Brechot P. Persistent hepatitis B virus infection in subjects without hepatitis B surface antigen: clinically significant or purely "occult"? Hepatology 2001;34:194–203.

[60] Kannangai R, Vivekanandan P, Netski D, et al. Liver enzyme flares and occult hepatitis B in persons with chronic hepatitis C infection. Clin Virol 2007;39:101–5.

[61] Cardoso C, Alves AL, Augusto F, et al. Occult hepatitis B infection in Portuguese patients with chronic hepatitis C liver disease: prevalence and clinical significance. Eur J Gastroenterol Hepatol 2013;25:142–6.

[62] Georgiadou SP, Zachou K, Rigopoulou E, et al. Occult hepatitis B virus infection in Greek patients with chronic hepatitis C and in patients with diverse nonviral hepatic diseases. J Viral Hepat 2004;11:358–65.

[63] Cacciola I, Pollicino T, Squadrito G, Cerenzia G, Orlando M, Raimondo G. Occult hepatitis B virus infection in patients with chronic hepatitis C liver disease. N Engl J Med 1999;341:22–6.

[64] Park JS, Saraf N, Dieterich DT. HBV plus HCV, HCV plus HIV, HBV plus HIV. Curr Gastroenterol Rep 2006;8:67–74.

[65] Fukuda R, Ishimura N, Hamamoto S, et al. Co-infection by serologically-silent hepatitis B virus may contribute to poor interferon response in patients with chronic hepatitis by down-regulation of type-I interferon receptor gene expression in the liver. J Med Virol 2001;63:220–7.

[66] De Maria N, Colantoni A, Friedlander L, et al. The impact of previous HBV infection on the course of chronic hepatitis C. Am J Gastroenterol 2000;95:3529–36.

[67] Chen LW, Chien RN, Yen CL, Chang JJ, Liu CJ, Lin CL. Therapeutic effects of pegylated interferon plus ribavirin in chronic hepatitis C patients with occult hepatitis B virus dual infection. J Gastroenterol Hepatol 2010;25:259–63.

[68] Weltman MD, Brotodihardjo A, Crewe EB, Farrell GC, Bilous M, Grierson JM, Liddle C. Coinfection with hepatitis B and C or B, C and delta viruses results in severe chronic liver disease and responds poorly to interferon-alpha treatment. J Viral Hepat 1995;2:39–45.

[69] Zignego AL, Fontana R, Puliti S, Barbagli S, Monti M, Careccia G, Giannelli F, et al. Relevance of inapparent coinfection by hepatitis B virus in alpha interferon-treated patients with hepatitis C virus chronic hepatitis. J Med Virol 1997;51:313–8.

[70] Sagnelli E, Coppola N, Scolastico C, Mogavero AR, Stanzione M, Filippini P, Felaco FM, et al. Isolated anti-HBc in chronic hepatitis C predicts a poor response to interferon treatment. J Med Virol 2001;65:681–7.

[71] Khattab E, Chemin I, Vuillermoz I, Vieux C, Mrani S, Guillaud O, Trepo C, et al. Analysis of HCV co-infection with occult hepatitis B virus in patients undergoing IFN therapy. J Clin Virol 2005;33:150–7.

[72] Villa E, Grottola A, Buttafoco P, Colantoni A, Bagni A, Ferretti I, Cremonini C, et al. High doses of alpha-interferon are required in chronic hepatitis due to coinfection with hepatitis B virus and hepatitis C virus: long term results of a prospective randomized trial. Am J Gastroenterol 2001;96:2973–7.

[73] Chuang WL, Dai CY, Chang WY, Lee LP, Lin ZY, Chen SC, Hsieh MY, et al. Viral interaction and responses in chronic hepatitis C and B coinfected patients with interferon-alpha plus ribavirin combination therapy. Antivir Ther 2005;10:125–33.

[74] Ye B, Shen J, Xu Y. Etiologic study on the relationship between HBV, HCV and HCC. Zhonghua Liu Xing Bing Xue Za Zhi 1994;15:131–4.

[75] Liaw YF. Hepatitis C virus superinfection in patients with chronic hepatitis B virus infection. J Gastroenterol 2002;37(Suppl. 13):65–8.

[76] Liaw YF, Yeh CT, Tsai SL. Impact of acute hepatitis B virus superinfection on chronic hepatitis C virus infection. Am J Gastroenterol 2000;95:2978–80.

[77] Wu JC, Chen CL, Hou MC, et al. Multiple viral infection as the most common cause of fulminant and subfulminant viral hepatitis in an area endemic for hepatitis B: application and limitations of the polymerase chain reaction. Hepatology 1994;19:836–40.

[78] Chu CM, Sheen IS, Liaw YF. The role of hepatitis C virus in fulminant viral hepatitis in an area with endemic hepatitis A and B. Gastroenterology 1994;107:189–95.

[79] Chu CJ, Lee SD. Hepatitis B virus/hepatitis C virus coinfection: epidemiology, clinical features, viral interactions and treatment. J Gastroenterol Hepatol 2008;23:512–20.

[80] Jamma S, Hussain G, Lau DT. Current concepts of HBV/HCV coinfection: coexistence, but not necessarily in harmony. Curr Hepat Rep 2010;9:260–9.

[81] Raimondo G, Brunetto MR, Pontisso P, Smedile A, Maina AM, Saitta C, et al. Longitudinal evaluation reveals a complex spectrum of virological profiles in hepatitis B virus/hepatitis C virus-coinfected patients. Hepatology 2006;43:100–7.

[82] Liu CJ, Chen PJ, Lai MY, Kao JH, Jeng YM, Chen DS. Ribavirin and interferon is effective for hepatitis C virus clearance in hepatitis B and C dually infected patients. Hepatology 2003;37:568–76.

[83] Liu CJ, Chuang WL, Lee CM, et al. Peg-interferon alfa-2a plus ribavirin for the treatment of dual chronic infection with hepatitis B and C viruses. Gastroenterology 2009;136:496–504.

[84] Yu ML, Lee CM, Chen CL, Taiwan Liver-Net Consortium, et al. Sustained hepatitis C virus clearance and increased hepatitis B surface antigen seroclearance in patients with dual chronic hepatitis C and B during posttreatment follow-up. Hepatology 2013;57:2135–42.

[85] Collins JM, Raphael KL, Terry C, Cartwright EJ, AnjanaIllai A, Anania FA, Farley MM. Hepatitis B virus reactivation during successful treatment of hepatitis C virus with Sofosbuvir and simeprevir. Clin Infect Dis 2015;61(8):1304–6.

[86] American Association for the Study of Liver Diseases/Infectious Diseases Society of American/International AIDS Society–USA. Recommendations for testing, managing, and treating hepatitis C. Available at: www.hcvguidelines.org; 2015.

[87] Brunetto MR, Cerenzia MT, Oliveri F, et al. Monitoring the natural course and response to therapy of chronic hepatitis B with an automated semi-quantitative assay for IgM anti-HBc. J Hepatol 1993;19:431–436.

[87a] Zacharakis G, Alzahrani J. Coinfection with hepatitis B and hepatitis C virus: new insights of treatment in the era of direct acting antivirals. Int J Adv Res 2017;5(5):1928–37.

[87b] Bersoff-Matcha SJ, Cao K, Jason M, Ajao A, Jones SC, Meyer T, Brinker A. Hepatitis B virus reactivation associated with direct-acting antiviral therapy for chronic hepatitis C virus: a review of cases reported to the U.S. Food and Drug Administration adverse event reporting system. Ann Intern Med 2017;166(11):792–8.

[88] US National Institutes of Health. ClinicalTrials.gov. 2015. https://clinicaltrials.gov/ct2/show/NCT02555943. First received September 20, 2015.

[89] Liu JY, Sheng YJ, Hu HD, et al. The influence of hepatitis B virus on antiviral treatment with interferon and ribavirin in Asian patients with hepatitis C virus/hepatitis B virus coinfection: a meta-analysis. Virol J 2012;9:186.

[90] Yalcin K, Degertekin H, Yildiz F, Kilinc N. A severe hepatitis flare in an HBV-HCV coinfected patient during combination therapy with A-interferon and ribavirin. J Gastroenterol 2003;38:796–800.

[91] Lampertico P, Agarwal K, Berg T, Buti M, Janssen HLA, Papatheodoridis G, Zoulim F, Tacke F. EASL 2017 Clinical practice guidelines on the management of hepatitis B virus infection. J Hepatol August 2017;67(2):370–98.

[92] Cheng HR, Liu CJ, Tseng TC, et al. Host genetic factors affecting spontaneous HBsAg seroclearance in chronic hepatitis B patients. PLoS One 2013;8:e53008.

[93] Marrone A, Zampino R, D'Onofrio M, Ricciotti R, Ruggiero G, Utili R. Combined interferon plus lamivudine treatment in young patients with dual HBV (HBeAg positive) and HCV chronic infection. J Hepatol 2004;41:1064–5.

[94] Coppola N, Stanzione M, Messina V, et al. Tolerability and efficacy of anti-HBV nucleos(t)ide analogues in HBV-DNA-positive cirrhotic patients with HBV/HCV dual infection. J Viral Hepat 2012;19:890–6.

[95] Omata M, Kanda T, Yu ML, et al. APASL consensus statements and management algorithms for hepatitis C virus infection. Hepatol Int 2012;6:409–35.

[96] Liaw YF, Kao JH, Piratvisuth T, et al. Asian-Pacific consensus statement on the management of chronic hepatitis B: a 2012 update. Hepatol Int 2012;6:531–61.

[97] Potthoff A, Berg T, Wedemeyer H. Late hepatitis B virus relapse in patients coinfected with hepatitis B virus and hepatitis C virus after antiviral treatment with pegylated interferon-a2b and ribavirin. Scand J Gastroenterol 2009;44:1487–90.

Chapter 4.5

Hepatitis C Infection in Patients With Hemoglobinopathies

Sanaa M. Kamal[1], Ahmed M. Fouad[2]

[1]*Ain Shams Faculty of Medicine, Cairo, Egypt;* [2]*American University of Lebanon, Beirut, Lebanon*

INTRODUCTION

Hereditary hemoglobin disorders cause a variety of syndromes, all with anemia as the common characteristic, and with a wide spectrum of clinical severity [1,2]. The most important clinically are those in whom anemia is so severe that life cannot be supported without regular blood transfusions. These include the beta-thalassemias, the compound beta and HbE thalassemias, and some forms of nondeletional alpha-thalassemias. They are caused by mutations affecting the production of α-globin chains and β-globin chains of the hemoglobin molecule. The severity of the anemia and its consequences depend on the molecular defects, which are involved in each affected individual. In addition to the thalassemia syndromes, there are phenotypically different syndromes, which are caused by variants of the hemoglobin molecule, mainly HbS and hemoglobin C, which cause sickle cell (SC) disease. In this chapter we discuss the importance of epidemiologic information in the management of these disorders [1–3] (Table 4.5.1).

Hepatitis C in Developing Countries. http://dx.doi.org/10.1016/B978-0-12-803233-6.00014-X
177

TABLE 4.5.1 Annual Number of Births With the Different Hemoglobin Disorders

Annual Births With Major Hemoglobin Disorders	
Beta-thalassemia major	22,989
HbE beta-thalassemia	19,128
HbH disease	9568
Hb Bart's hydrops (α^0/α^0)	5183
Hemoglobin SS disease	217,331
S beta-thalassemia	11,074
Sickle cell disease	54,736

Adapted from Weatherall DJ. The inherited diseases of haemoglobin are an emerging global health burden. Blood June 3, 2010;115(22):4331–36.

The thalassemia syndromes, particularly those requiring multiple blood transfusions, represent a serious burden on health services, a problem which may be increasing on a global scale [1–4]. Even milder syndromes, such as thalassemia intermedia or nontransfusion-dependent thalassemia, require careful follow-up because complications are expected over time, in the natural course of the disease. This is also true of the SC syndromes. The need for lifelong follow-up and care and the occurrence of complications affecting major organs such as liver, heart, and endocrine glands create the need for organized expert services and also the need for major resources in terms of essential drugs and donated blood for transfusions [1,2]. As chelation therapy with new drugs seems to prevent cardiac damage and improve survival, chronic liver disease has emerged as a critical clinical issue in this setting.

Hepatitis C virus (HCV) infection is a leading cause of chronic liver disease in patients with multiple transfusions, who represent a population at high risk of acquiring HCV as before 1992 there was no blood screening for HCV. As a result, the prevalence of antibodies to HCV in patients with thalassemia varies depending on the population studied between 12% and 85% [3,4]. The "hit" to the liver is at least dual: high frequency of chronic viral infections, especially HCV, and secondary hemochromatosis of the liver because of multiple transfusions and dyserythropoiesis. Furthermore, chronic hepatitis C (CHC) and iron overload were proved to be independent risk factors for liver fibrosis progression and their concomitant presence results in a striking increase of the risk [4].

The treatment of HCV in patients with thalassemia has been interferon-α (IFN-α) with or without ribavirin (RBV) with sustained virologic response (SVR) rates of 40% to 60% [4,5]. Direct-acting antivirals (DAAs) caused a breakthrough in the treatment of CHC regarding extremely high SVR rates, better compliance, and significantly lower side effects. Furthermore, several oral non-RBV–containing

DAAs are available. To date, very little data are available on the results of DAAs in patients with hemoglobinopathies and CHC. However, such regimens seem promising in the management of HCV in patients with hemoglobinopathies.

HEMOGLOBINOPATHIES IN DEVELOPING COUNTRIES

SC Disease

SC disease is a common, multisystem, monogenic disorder. SC disease is a disorder that is caused by a single gene mutation [2]. SC disease is characterized by the presence of abnormal erythrocytes damaged by HbS, this variant of normal adult hemoglobin is inherited either from both parents (homozygosity for the HbS gene) or from one parent, along with another hemoglobin variant, such as HbC, or with beta-thalassemia (compound heterozygosity). When deoxygenated, HbS polymerizes, damaging the erythrocyte and causing it to lose cations and water. These damaged cells have abnormalities in the expression of adhesion molecules, resulting in hemolytic anemia and a likelihood of blocked small blood vessels, which in turn cause vaso-occlusion. Vaso-occlusion typically causes acute complications, including ischemic damage to tissues, resulting in severe pain or organ failure [4].

Epidemiology of SC Disease

There is no reliable global estimates for the incidence of SC anemia; however, it is estimated that every year approximately 300,000 infants are born with SC anemia and that this number could rise to 400,000 by 2050 [5,6]. The vast majority of these births occur in three countries: Nigeria, the Democratic Republic of the Congo, and India [2,7–9]. In several sections of Africa, the prevalence of SC trait (heterozygosity) is as high as 30% [2,8]. Pockets of SC disease exist in the Mediterranean basin, the Middle East, and India because of the remarkable level of protection that the SC trait (i.e., heterozygosity for the SC mutation in *HBB*) provides against severe malaria [2,7]. By the use of haplotype analysis, it has been determined that the SC gene arose at least twice, once in Africa and once in India or the Middle East [4]. Hemoglobin SC disease is more restricted to parts of west and north Africa, whereas HbS beta-thalassemia occurs in localized parts of sub-Saharan Africa and sporadically throughout the Middle East and Indian subcontinent. Unlike some forms of α thalassemia and HbE, the gene frequency for HbS rarely rises much above 20% to 25% of a particular population, although there are occasional exceptions [2,7,8]. HbS is transmitted as an autosomal codominant characteristic. The male-to-female ratio is 1:1. No sex predilection exists, because SC anemia is not an X-linked disease [6,7].

A strong geographical link between the highest HbS allele frequencies and high malaria endemicity has been reported particularly in Africa. The gradual increase in HbS allele frequencies from epidemic areas to holoendemic areas in Africa is consistent with the hypothesis that malaria protection by HbS involves the enhancement of not only innate but also acquired immunity to *Plasmodium falciparum* [7]. Despite the presence of large malarious areas, HbS is absent in the Americas and in large parts of Asia [7,10,11]. The complex social structure and the predominance of *Plasmodium vivax* are also considered as likely to contribute to the unresolved geographical relationship in India [12].

Beta-Thalassemia

Beta-thalassemia syndromes are a group of hereditary disorders characterized by a genetic deficiency in the synthesis of beta-globin chains. In the homozygous state, beta-thalassemia (i.e., thalassemia major) causes severe, transfusion-dependent anemia. In the heterozygous state, the beta-thalassemia trait (i.e., thalassemia minor) causes mild to moderate microcytic anemia. Patients in whom the clinical severity of the disease lies between that of thalassemia major and thalassemia minor are categorized as having thalassemia intermedia. Several different genotypes are associated with thalassemia, such as HbC/beta-thalassemia, HbE/beta-thalassemia, and HbS/beta-thalassemia, which is more similar to SC disease than to thalassemia major or intermedia [13].

Epidemiology of Beta-Thalassemia

Beta-thalassemia is prevalent in Mediterranean countries, the Middle East, Central Asia, India, Southern China, and the Far East and countries along the north coast of Africa and in South America. The highest carrier frequency is reported in Cyprus (14%), Sardinia (10.3%), and Southeast Asia [14]. Population migration and intermarriage between different ethnic groups has introduced thalassemia in almost every country of the world, including Northern Europe where thalassemia was previously absent. It has been estimated that about 1.5% of the global population (80–90 million people) are carriers of beta-thalassemia, with about 60,000 symptomatic individuals born annually, the great majority in the developing world. The total annual incidence of symptomatic individuals is estimated at 1 in 100,000 throughout the world [14,15]. However, accurate data on carrier rates in many populations are lacking, particularly in areas of the world known or expected to be heavily affected.

It is estimated that 1.5% of the world's population are carriers of beta-thalassemia. Southeast Asia accounts for about 50% of the world's population [15]. Beta-thalassemia is particularly prevalent among the Mediterranean populations particularly in Greece, Italy, Egypt, and Lebanon, but it is less common at the western end of the Mediterranean. Beta-thalassemia is reported to be between 3% and 7% in most of North Africa. Beta-thalassemia is also common in the Middle East and west Asia [16–19] (Table 4.5.2).

Alpha-Thalassemia

Alpha-thalassemia is one of the hemoglobin genetic abnormalities, which is caused by the reduced or absent production of the alpha globin chains. Alpha-thalassemia is prevalent in tropical and subtropical world regions where malaria

TABLE 4.5.2 Common Types of Beta-Thalassemia: Severity and Ethnic Distribution

Population	β-Gene Mutation	Severity
Indian	−619 del	β^0
Mediterranean	−101 C→T	β^{++}
Black	−88 C→T	β^{++}
Mediterranean; African	−87 C→G	β^{++}
Japanese	−31 A→G	β^{++}
African	−29 A→G	β^{++}
Southeast Asian	−28 A→C	β^{++}
Mediterranean; Asian Indian	IVS1-nt1 G→A	β^0
East Asian; Asian Indian	IVS1-nt5 G→C	β^0
Mediterranean	IVS1-nt6 T→C	$\beta^{+/++}$
Mediterranean	IVS1-nt110 G→A	β^+
Chinese	IVS2-nt654 C→T	β^+
Mediterranean	IVS2-nt745 C→G	β^+
Mediterranean	Codon 39 C→T	β^0
Mediterranean	Codon 5 −CT	β^0
Mediterranean; African-American	Codon 6 −A	β^0
Southeast Asian	Codon 41/42 −TTCT	β^0
African-American	AATAAA to AACAAA	β^{++}
Mediterranean	AATAAA to AATGAA	β^{++}
Mediterranean	Codon 27 G→T Hb (Hb Knossos)	β^{++}
Southeast Asian	Codon 79 G>A (Hb E)	β^{++}
Malaysia	Codon 19 G>A (Hb Malay)	

β^0, complete absence of beta globin on the affected allele; β^+, residual production of beta globin (around 10%); β^{++}, very mild reduction in beta globin production.
Adapted from the Thalassemia International Federation; Prati D, Zanella A, Farma E, et al. A multicenter prospective study on the risk of acquiring liver disease in anti-hepatitis C virus negative patients affected from homozygous beta-thalassemia. Blood 1998;92:3460–64.

was and still is epidemic, but as a consequence of the recent massive population migrations, alpha-thalassemia has been introduced to North America, North Europe, and Australia. Alpha-thalassemia is very heterogeneous at a clinical and molecular level. Four clinical conditions of increased severity are recognized: the silent carrier state, the alpha-thalassemia trait, the intermediate form of hemoglobin H disease, and the hemoglobin Bart hydrops fetalis syndrome that is lethal in utero or soon after birth [14,15].

Alpha-thalassemia is caused most frequently by deletions involving one or both alpha globin genes and less commonly by nondeletional defects. A large number of alpha-thalassemia alleles have been described, and their interaction results in the wide spectrum of hematologic and clinical phenotypes. Carriers of alpha-thalassemia do not need any treatment. Usually, patients with hemoglobin H disease are clinically well and survive without any treatment, but occasional red blood cell transfusions may be needed if the hemoglobin level suddenly drops because of hemolytic or aplastic crisis likely because of viral infections [15].

HEPATITIS C INFECTION IN PATIENTS WITH HEMOGLOBINOPATHIES

Epidemiology

Hepatitis C is a major cause of liver disease in patients with hemoglobinopathies. Before the implementation of blood screening for HCV in 1991, blood transfusion was the most frequent route for acquiring HCV in patients with hemoglobinopathies. In developed countries and many developing countries, HCV is no longer a threat in blood supply. However, blood safety is still questionable in some resource-limited countries because of poor facilities [16,17]. Other routes of transmission, such as injection drug use, iatrogenic exposure, health-related practices, traditional practices, occupational transmission, and sexual intercourse, contribute to the transmission of HCV among patients with various forms of hemoglobinopathies in developing countries [16].

Liver Disease Patterns in Patients With Thalassemia With HCV Infection

Studies showed that 80% to 90% of patients with hemoglobinopathies who receive multiple transfusions and those who received blood transfusions had HCV antibodies [18–27]. The rate of chronic HCV progression to cirrhosis in patients with thalassemia is highly variable and is influenced by posttransfusion as an independent risk factor for liver fibrosis progression. Moreover, hepatocellular carcinoma is becoming frequent with the aging population of patients with thalassemia [28,29].

Patients with HCV receiving multiple transfusions are subjected to more liver disease resulting from the combined effect of iron accumulation and toxicity, along with HCV infection, both of which have synergistic effects [30,31]. Iron overload has been shown by some studies to be an independent risk factor for liver fibrosis progression in those patients. In HCV, mono-infection ranges from 5% to 25% over periods of 25–30 years [32,33], whereas HCV-related cirrhotics are at risk of hepatic decompensation (30% over 10 years) and hepatocellular carcinoma (1%–3% per year) [32,33]. In transfusion-associated HCV infection, iron overload seems to confer to severe liver fibrosis and morbidity in the absence of HCV infection [34,35]. However, some studies [30–33] showed that adult patients with beta-thalassemia did not have marked increase in liver disease progression rates, despite the HCV and iron overload comorbidity.

Treatment of HCV in Patients With Hemoglobinopathies

Interferon-Based Regimen

For more than a decade, the treatment of chronic C infection in patients with hemoglobinopathies has been pegylated-IFN (Peg-IFN) with or without RBV. In many centers, RBV was contraindicated in this subset of patients, because of the risk of induced hemolytic anemia. Later, RBV was approved for these patients. Thus, treatment schedules became similar to those used in the general population. However, the SVR rates because of a combination of Peg-IFN and RBV led to SVR rates were highly dependent on viral genotype and stage of liver fibrosis with the best results in patients infected with genotypes 2 and 3. Presence of advanced liver fibrosis or cirrhosis had a negative impact on the overall SVR rates in patients with chronic HCV without and with hemoglobinopathies.

Studies on the efficacy of interferon-based regimen in patients with TM yielded heterogeneous results because of the use of different IFN regimens (IFN vs. Peg-IFN, RBV presence vs. absence, IFN/RBV doses, treatment duration) and the limited number of patients included. Most of the studies include patients with TM treated with IFN monotherapy. In an early study, 12 patients with TM were treated with IFN-a 2b monotherapy for 26 weeks; this therapeutic approach was not successful because 25% of those treated withdrew from the trial because of adverse events (AEs) [36]. Other studies evaluating the response of patients with HCV and thalassemia to IFN as monotherapy then followed [36–41]. Patients treated with IFN/Peg-IFN as monotherapy achieved overall SVR rates ranging between 26% and 80% according to HCV genotype and liver disease status (Table 4.5.3).

The addition of RBV to IFN/Peg-IFN led to an increase in the overall SVR rates, which ranged from 45% in those patients treated with IFN/RBV to 70% in those treated with Peg-IFN/RBV; however, no differences in treatment efficacy emerged between patients with TM treated with Peg-IFN or IFN (Table 4.5.3). HCV genotypes, absence of fibrosis and steatosis, low viral load, no previous

TABLE 4.5.3 Overall SVR Rates in Different Anti-HCV Therapeutic Regimen in Patients With Hemoglobinopathies

Study (Year)	Hemoglo-binopathy	Treatment Regimen	# Patients	SVR Rates (Overall)
Di Marco et al. [36]	TM	IFN	12	NA
Donohue et al. [37]	TM	IFN	12	NA
Clemente et al. [38]	TM	IFN	51	37%
Di Marco et al. [39]	TM	IFN	70	40%
Telfer et al. [40]	TM	IFN	11	45%
Pizzarelli et al. [41]	TM	IFN	51	NA
Spiliopoulou et al. [45]	TM	IFN	13	77%
Sievert et al. [46]	TM	IFN	28	64%
Li et al. [47]	TM	IFN + RBV	18	72%
Artan et al. [48]	TM	IFN	10	80%
Syriopoulou et al. [49]	TM	IFN	89	52%
Inati et al. [50]	TM	Overall	20	45%
		Peg-IFN	12	33%
		Peg-IFN + RBV	8	63%
Harmatz et al. [51]	TM	IFN + RBV	21	33%
Kalantari et al. [52]	TM	Overall	32	44%
		IFN	18	56%
		IFN + RBV	14	29%
Sood et al. [53]	TM	Overall	40	55%
		Peg-IFN	20	40%
		Peg-IFN + RBV	20	70%

TABLE 4.5.3 Overall SVR Rates in Different Anti-HCV Therapeutic Regimen in Patients With Hemoglobinopathies—Cont'd

Study (Year)	Hemoglo-binopathy	Treatment Regimen	# Patients	SVR Rates (Overall)
Jafroodi et al. [54]	TM	IFN+RBV	30	50%
Di Marco et al. [55]	TM	IFN	114	40%
Vafiadis et al. [56]	TM	Overall	48	31%
		IFN	29	34%
		Peg-IFN	19	26%
Kamal et al. [57]	TM	Overall		55%
		Peg-IFN		46%
		Peg-IFN+RBV		64%
Swaim et al. [58]	SC disease	IFN+RBV	2	100%
Ancel et al. [59]	SC disease	Peg-IFN+RBV	6	50%
Ayyub et al. [60]	SC disease	Peg-IFN+RBV	8	63%
Issa [61]	SC disease	Peg-IFN+RBV	52	71%

HCV, hepatitis C virus; *IFN*, interferon; *Peg*, pegylated; *RBV*, ribavirin; *SC*, Sickle cell; *SVR*, sustained virologic response; *TM*, thalassemia.

treatments, and some genetic features were shown as the strongest predictors of SVR [16]. Most of the patients included in these studies were IFN naive.

Treatment of Patients With Thalassemia With Chronic HCV in the Era of DAAs

Few studies evaluated the efficacy of DAAs in different forms of hemoglobinopathies [42–44] (Table 4.5.4). Such studies showed high SVR rates with few AEs. According the American Association for the Study of Liver Diseases (AASLD), Infectious Diseases Society of America (IDSA), and European Association of Study of Liver Diseases (EASL) guidelines, the indications for HCV therapy are the same in patients with and without hemoglobinopathies. Patients with hemoglobinopathies should be treated with an IFN-free regimen, without RBV.

TABLE 4.5.4 Treatment of HCV in Patients With Hemoglobinopathies in the Era of Direct-Acting Antivirals

Study (Year)	Hemoglobinopathy	Treatment Regimen	# Patients	Sustained virologic response Rates (Overall)
Sinakos et al. [62]	TM	• Sofosbuvir (SOF) + ribavirin (RBV) • SOF + simeprevir ± RBV • SOF + daclatasvir ± RBV • Ledipasvir/SOF ± RBV • Ombitasvir/paritaprevir-ritonavir + dasabuvir ± RBV	61	90%
Hezode et al. [42]	TM	Elbasvir/grazoprevir (EBR/GZR)	159	97.6%
	Sickle cell disease			94.7%
Köklü et al. [43]	TM	Daclatasvir and asunaprevir	1	100%
Papadopoulos et al. [44]	HbS Beta 0-Thalassemia	SOF + simeprevir	1	100%

TM, thalassemia.

The anti-HCV regimens that can be used in patients with hemoglobinopathies are the same as in patients without hemoglobinopathies. RBV-containing regimens are better excluded. However, when the use of RBV is needed, careful monitoring is recommended, and blood transfusion support may be required.

ABBREVIATIONS

IFN Interferon
Peg-IFN Pegylated interferon
RBV Ribavirin
SC disease Sickle cell disease
SOF Sofosbuvir
TM Thalassemia

REFERENCES

[1] Modell B, Darlison M. Global epidemiology of haemoglobin disorders and derived service indicators. Bull World Health Organ 2008;86(6):480–7.

[2] Weatherall DJ. The inherited diseases of haemoglobin are an emerging global health burden. Blood June 3, 2010;115(22):4331–6.

[3] Alwan A, Modell B. Recommendations for introducing genetics services in developing countries. Nat Rev Genet 2003;4(1):61–8.

[4] Wailoo K. Sickle cell disease – a history of progress and peril. N Engl J Med March 2, 2017;376(9):805–7.

[5] Piel FB, Patil AP, Howes RE, et al. Global epidemiology of sickle haemoglobin in neonates: a contemporary geostatistical model-based map and population estimates. Lancet 2013;381:142–51.

[6] Piel FB, Hay SI, Gupta S, Weatherall DJ, Williams TN. Global burden of sickle cell anaemia in children under five, 2010–2050: modelling based on demographics, excess mortality, and interventions. PLoS Med 2013;10:e1001484.

[7] Piel FB, Patil AP, Howes RE, et al. Global distribution of the sickle cell gene and geographical confirmation of the malaria hypothesis. Nat Commun 2010;1:104.

[8] Grosse SD, Odame I, Atrash HK, Amendah DD, Piel FB, Williams TN. Sickle cell disease in Africa: a neglected cause of early childhood mortality. Am J Prev Med 2011;41(Suppl. 4):S398–405.

[9] Kyu HH, Pinho C, Wagner JA, et al. Global and national burden of diseases and injuries among children and adolescents between 1990 and 2013: findings from the Global Burden of Disease 2013 study. JAMA Pediatr 2016;170:267–87.

[10] Modiano D, et al. Haemoglobin S and haemoglobin C: 'quick but costly' versus 'slow but gratis' genetic adaptations to *Plasmodium falciparum* malaria. Hum Mol Genet 2008;17:789–99.

[11] Gouagna LC, et al. Genetic variation in human HBB is associated with *Plasmodium falciparum* transmission. Nat Genet 2010;42:328–31.

[12] Guerra CA, et al. the international limits and population at risk of *Plasmodium vivax* transmission in 2009. PLoS Negl Trop Dis 2010;4:e774.

[13] Borgna-Pignatti C, Galanello R. Thalassemias and related disorders: quantitative disorders of hemoglobin synthesis. In: Wintrobe's Clinical Hematology. vol. 42. Philadelphia: Lippincott Williams & Wilkins; 2004. p. 1319–65. 11.

[14] Piel FB, Weatherall DJ. The α-thalassemias. N Engl J Med November 13, 2014;371(20):1908–16.

[15] Higgs DR, Weatherall DJ. The alpha thalassaemias. Cell Mol Life Sci April 2009;66(7):1154–62.

[16] Alter M. Epidemiology of hepatitis C virus infection. World J Gastroenterol May 7, 2007;
13(17):2436–41.

[17] Luban NLC. Transfusion safety: where are we today? Ann N Y Acad Sci 2005:1054.

[18] Colombo M, Oldani S, Donato MF, Borzio M, Santese R, Roffi L, Viganó P, Cargnel A. A multicenter, prospective study of posttransfusion hepatitis in milan. Hepatology 1987;7:709–12.

[19] Velati C, Romanò L, Fomiatti L, Baruffi L, Zanetti AR, SIMTI Research Group. Impact of
nucleic acid testing for hepatitis B virus, hepatitis C virus, and human immunodeficiency virus
on the safety of blood supply in Italy: a 6-year survey. Transfusion 2008;48:2205–13.

[20] Prati D, Zanella A, Farma E, et al. A multicenter prospective study on the risk of acquiring liver
disease in anti-hepatitis C virus negative patients affected from homozygous beta-thalassemia.
Blood 1998;92:3460–4.

[21] Mahmoud RA, El-Mazary AA, Khodeary A. Seroprevalence of hepatitis C, hepatitis B, cytomegalovirus, and human immunodeficiency viruses in multitransfused thalassemic children in
Upper Egypt. Adv Hematol 2016;2016:9032627.

[22] Kissou SA, Koura M, Sawadogo A, Ouédraogo AS, Traoré H, Kamboulé E, Zogona WWF,
Nacro B. Serological markers of viral hepatitis B and C in children with sickle cell disease
monitored in the Pediatrics Department at the University Hospital of Bobo-Dioulasso (Burkina
Faso). Bull Soc Pathol Exot April 17, 2017.

[23] Aminianfar M, Khani F, Ghasemzadeh I. Evaluation of hepatitis C, hepatitis B, and HIV virus
serology pandemic in thalassemia patients of Shahid Mohammadi Hospital of Bandar Abbas,
Iran. Electron Phys March 25, 2017;9(3):4014–9.

[24] Chakravarti A, Verma V. Distribution of hepatitis C virus genotypes in beta-thalassaemic
patients from Northern India. Transfus Med December 2006;16(6):433–8.

[25] Namasopo SO, Ndugwa C, Tumwine JK. Hepatitis C and blood transfusion among children attending the sickle cell clinic at Mulago Hospital, Uganda. Afr Health Sci June 2013;13(2):255–60.

[26] Diarra AB, Guindo A, Kouriba B, Dorie A, Diabaté DT, Diawara SI, Fané B, Touré BA,
Traoré A, Gulbis B, Diallo DA. Sickle cell anemia and transfusion safety in Bamako, Mali.
Seroprevalence of HIV, HBV and HCV infections and alloimmunization belonged to Rh and
Kell systems in sickle cell anemia patients. Transfus Clin Biol 2013;20(5–6):476–81.

[27] Sack FN, Noah Noah D, Zouhaïratou H, Mbanya D. Prevalence of HBsAg and anti-HCV antibodies in homozygous sickle cell patients at Yaounde Central Hospital. Pan Afr Med J 2013;14:40.

[28] Papatheodoridis G, Thomas HC, Golna C, Bernardi M, Carballo M, Cornberg M, Dalekos G,
Degertekin B, Dourakis S, Flisiak R, Goldberg D, Gore C, Goulis I, Hadziyannis S, Kalamitsis
G, Kanavos P, Kautz A, Koskinas I, Leite BR, Malliori M, Manolakopoulos S, Maticic M,
Papaevangelou V, Pirona A, Prati D, Raptopoulou-Gigi M, Reic T, Robaeys G, Schatz E,
Souliotis K, Tountas Y, Wiktor S, Wilson D, Yfantopoulos J, Hatzakis A. Addressing barriers to
the prevention, diagnosis and treatment of hepatitis B and C in the face of persisting fiscal con-
straints in Europe: report from a high level conference. J Viral Hepat 2016;23(Suppl. 1):1–12.

[29] Di Marco V, Capra M, Gagliardotto F, Borsellino Z, Cabibi D, Barbaria F, Ferraro D, Cuccia
L, Ruffo GB, Bronte F, Di Stefano R, Almasio PL, Craxì A. Liver disease in chelated transfusion dependent thalassemics: the role of iron overload and chronic hepatitis C. Haematologica
2008;93:1243–6.

[30] Kountouras D, Tsagarakis NJ, Fatourou E, Dalagiorgos E, Chrysanthos N, Berdoussi H, Vgontza
N, Karagiorga M, Lagiandreou A, Kaligeros K, Voskaridou E, Roussou P, Diamanti-Kandarakis
E, Koskinas J. Liver disease in adult transfusion-dependent beta-thalassaemic patients: investi-
gating the role of iron overload and chronic HCV infection. Liver Int March 2013;33(3):420–7.

[31] Triantos C, Kourakli A, Kalafateli M, Giannakopoulou D, Koukias N, Thomopoulos K, Lampropoulou P, Bartzavali C, Fragopanagou H, Kagadis GC, Christofidou M, Tsamandas A, Nikolopoulou V, Karakantza M, Labropoulou-Karatza C. Hepatitis C in patients with β-thalassemia major. A single-centre experience. Ann Hematol June 2013;92(6):739–46.

[32] Lai M, Origa R, Danjou F, Gian B, Vacquer S, Franco Anni F, Corrias C, Farci P, Congiu G, Galanello R. Natural history of hepatitis C in thalassemia major: a long-term prospective study. Eur J Haematol 2013;90:501–7.

[33] Liang TJ, Rehermann B, Seeff LB, Hoofnagle JH. Pathogenesis, natural history, treatment, and prevention of hepatitis C.. Ann Intern Med 2000;132:296–305.

[34] Hershko C. Pathogenesis and management of iron toxicity in thalassemia. Ann N Y Acad Sci 2010;1202:1–9.

[35] Angelucci E, Muretto P, Nicolucci A, Baronciani D, Erer B, Gaziev J, Ripalti M, Sodani P, Tomassoni S, Visani G, Lucarelli G. Effects of iron overload and hepatitis C virus positivity in determining progression of liver fibrosis in thalassemia following bone marrow transplantation. Blood 2002;100:17–21.

[36] Di Marco V, Lo Iacono O, Capra M, et al. Interferon treatment of chronic hepatitis C in young patients with homozygous beta-thalassemia. Hematologica 1992;77:502–6.

[37] Donohue SM, Wonke B, Hoffbrand AV, et al. Alpha interferon in the treatment of chronic hepatitis C infection in thalassemia major. Br J Hematol 1993;83:491–7.

[38] Clemente MG, Congia M, Lai ME, et al. Effect of iron overload on the response to recombinant interferon-alfa treatment in transfusion-dependent patients with thalassemia major and chronic hepatitis C. J Pediatr 1994;125:123–8.

[39] Di Marco V, Lo Iacono O, Almasio P, et al. Long-term efficacy of alpha-interferon in beta-thalassemics with chronic hepatitis C. Blood 1997;90:2207–12.

[40] Telfer PT, Garson JA, Whitby K, et al. Combination therapy with interferon alpha and ribavirin for chronic hepatitis C virus infection in thalassemic patients. Br J Hematol 1997;98:850–5.

[41] Pizzarelli G, Di Gregorio F, Romeo MA, et al. Interferon-α therapy in Sicilian and Sardinian polytransfused thalassemic patients with chronic hepatitis C. Biodrugs 1999;12:55–63.

[42] Hézode C, Colombo M, Bourlière M, Spengler U, Ben-Ari Z, Strasser SI, Lee WM, Morgan L, Qiu J, Hwang P, Robertson M, Nguyen BY, Barr E, Wahl J, Haber B, Chase R, Talwani R, Di Marco V, C-EDGE IBLD Study Investigators. Elbasvir/Grazoprevir for patients with hepatitis C virus infection and inherited blood disorders: a phase III study. Hepatology March 3, 2017.

[43] Köklü H, Köklü S, Ozturk O, Aksoy EK, Tseveldorj N. Successful treatment of chronic hepatitis C virus infection with daclatasvir and asunaprevir in a transfusion dependent thalassaemia patient. Acta Gastroenterol Belg September–December 2016;79(4):501–2.

[44] Papadopoulos N, Deutsch M, Georgalas A, Poulakidas H, Karnesis L. Simeprevir and sofosbuvir combination treatment in a patient with HCV cirrhosis and HbS beta 0-thalassemia: efficacy and safety despite baseline hyperbilirubinemia. Case Rep Hematol 2016;2016:7635128.

[45] Spiliopoulou I, Repanti M, Katinakis S, et al. Response to interferon alfa-2b therapy in multitransfused children with beta-thalassemia and chronic hepatitis C. Eur J Clin Microbiol Dis 1999;18:709–15.

[46] Sievert W, Pianko S, Warner S, Bowden S, Simpson I, Bowden D, Locarnini S. Hepatic iron overload does not prevent a sustained virological response to interferon-alpha therapy: a long term follow-up study in hepatitis C-infected patients with beta thalassemia major. Am J Gastroenterol April 2002;97(4):982–7.

[47] Li CK, Chan PKS, Ling SC, et al. Interferon and ribavirin as frontline treatment for chronic hepatitis C infection in thalassemia major. Br J Hematol 2002;117:755–8.

[48] Artan R, Akcam M, Yilmaz A, Kocacik D. Interferon alpha monotherapy for chronic hepatitis C viral infection in thalassemics and hemodialysis patients. J Chemother 2005;6:651–5.

[49] Syriopoulou V, Daikos GL, Kostaridou SL, et al. Sustained response to interferon alpha-2a in thalassemic patients with chronic hepatitis C. A prospective 8-year follow-up study. Hematologica 2005;90:129–31.

[50] Inati A, Taher A, Ghorra S, et al. Efficacy and tolerability of peginterferon alpha-2a with or without ribavirin in thalassemia major patients with chronic hepatitis C virus infection. Br J Hematol 2005;130:644–6.

[51] Harmatz P, Jonas MM, Kwiatkowski JL, et al. Safety and efficacy of pegylated interferon a-2a and ribavirin for the treatment of hepatitis C in patients with thalassemia. Hematologica 2008;93:1247–51.

[52] Kalantari H, Rad N. Efficacy of interferon alpha-2b with or without ribavirin in thalassemia major patients with chronic hepatitis C virus infection: a randomized, double blind, controlled, parallel group trial. J Res Med Sc 2010;15:310–6.

[53] Sood A, Sobti P, Midha V, et al. Efficacy and safety of pegylated IFN alfa 2b alone or in combination with ribavirin in thalassemia major with chronic hepatitis C.. Indian J Gastroenterol 2010;29:62–5.

[54] Jafroodi M, Asadi R, Heydarzadeh A, Besharati S. Effect of hepatic iron concentration and viral factors in chronic hepatitis C-infected patients with thalassemia major, treated with interferon and ribavirin. Int J Gen Med 2011;4:529–33.

[55] Di Marco V, Bronte F, Calvaruso V, et al. IL28B polymorphisms influence stage of fibrosis and spontaneous or interferon-induced viral clearance in thalassemia patients with hepatitis C virus infection. Hematologica 2012;97:679–86.

[56] Vafiadis I, Trilianos P, Vlachogiannakos J, et al. Efficacy and safety of interferon-based therapy in the treatment of adult thalassemic patients with chronic hepatitis C: a 12 years audit. Ann Hepatol 2013;12:364–70.

[57] Kamal SM, Fouly A, Mohamed S, et al. Peginterferon-alpha2b therapy with and without ribavirin in patients with thalassemia: a randomized study. J Hepatol 2006;44(Suppl. 2):S217.

[58] Swaim MW, Agarwal S, Rosse WF. Successful treatment of hepatitis C in sickle-cell disease. Ann Intern Med 2000;133(9):750–1.

[59] Ancel DB, Amiot X, Chaslin-Ferbus D, et al. Treatment of chronic hepatitis C in sickle cell disease and thalassemic patients with interferon and ribavirin. Eur J Gastroenterol Hepatol 2009;21:726–9.

[60] Ayyub MA, El-Moursy SA, Khazindar AM, Abbas FA. Successful treatment of chronic hepatitis C virus infection with peg-interferon alpha-2a and ribavirin in patients with sickle-cell disease. Saudi Med J 2009;30:712–6.

[61] Issa H. Safety of pegylated interferon and ribavirin therapy for chronic hepatitis C in patients with sickle cell anemia. World J Hepatol May 27, 2010;2(5):180–4.

[62] Sinakos E, Kountouras D, Koskinas J, Zachou K, Karatapanis S, Triantos C, Vassiliadis T, Goulis I, Kourakli A, Vlachaki E, Toli B, Tampaki M, Arvaniti P, Tsiaoussis G, Bellou A, Kattamis A, Maragkos K, Petropoulou F, Dalekos GN, Akriviadis E, Papatheodoridis GV. Treatment of chronic hepatitis C with direct-acting antivirals in patients with β-thalassaemia major and advanced liver disease. Br J Haematol April 25, 2017.

FURTHER READING

[1] Seeff LB. Natural history of chronic hepatitis C. Hepatology 2002;36:S35–46.

Chapter 4.6

Hepatitis C in Patients on Hemodialysis

Tamer A. Hafez
American University, Cairo, Egypt

Chapter Outline

INTRODUCTION

Chronic renal diseases (CRDs) in developing countries are predominantly glomerulonephritis and hypertension, although type 2 diabetes is also becoming a significant cause. In developing countries, patients with CRD are generally younger than those in the developed world, and there is a significant male preponderance. Hemodialysis (HD) is the most commonly used modality of renal replacement therapy (RRT) in developing countries. Most patients are managed by HD, with peritoneal dialysis and kidney transplantation (KT) being available in only few countries in the region. Government funding and support for dialysis are often unavailable and, when available, often with restrictions. There is a marked deficiency of trained manpower to treat CRD, and many countries have a limited number of units, which are often ill-equipped to deal adequately with the number of patients who require end-stage renal disease (ESRD) care in the resource-limited countries [1–3]. In developing countries such as those countries in sub-Saharan Africa, HD is the most common modality of RRT [3]. Peritoneal dialysis and KT are largely uncommon because of the extremely high costs, lack of facilities and manpower, and the predominantly urban location

Hepatitis C in Developing Countries. http://dx.doi.org/10.1016/B978-0-12-803233-6.00015-1
191

of the renal care centers [3,4]. As a result of financial constraints, dialyzers are frequently reused and water is inadequately treated. A majority of patients decrease session frequency or discontinue the program because of financial constraints [3–5]. Thus, patients with ESRD undergoing HD are exposed to numerous health risks [4–6].

Hepatitis C virus (HCV) infection is a significant cause of morbidity and mortality in patients with HD and kidney transplant recipients specifically in developing countries. In developing countries, the prevalence of anti-HCV sero-positivity among patients on maintenance HD ranges between 25% and 60%. Patients on HD are at a high risk for HCV, with frequency of infection several times higher than that in nonuremic patients [7–10].

THE IMPACT OF HCV ON KIDNEY DISEASE

A meta-analysis [11] revealed that HD can negatively modify the course of HCV infection. The estimated relative risk of liver-related mortality in patients who are anti-HCV positive and on HD was 1.57 times [95% confidence interval (CI): 1.33–1.86; $P < .001$] than that observed for the anti-HCV–negative counterparts. Thus, the presence of anti-HCV antibodies in patients with ESRD on HD is an independent risk factor for death mostly because of the increased risk of cirrhosis and hepatocellular carcinoma. The impact of HCV on KT is controversial. Some studies showed that KT improves the long-term survival of patients with ESRD and with HCV infection [12,13], whereas other studies showed that HCV infection threatens the success of KT [14]. However, a meta-analysis [15] revealed that the survival of HCV-infected renal transplant recipients is better than that for patients with HCV infection undergoing HD who are on transplant waiting lists. This survival advantage may reflect systemic effects of well-functioning renal allografts that is clearing uremic toxins and may also reflect reduced inflammatory responses and oxidative stress. HCV-related deterioration of renal transplant recipients may be linked to the immunosuppressive treatment that is required after KT. This can result in flares of HCV infection and can increase liver- and kidney-related morbidity and mortality from conditions such as cirrhosis, hepatocellular carcinoma, transplant glomerulopathy, and graft dysfunction [15].

PREVENTION OF HCV TRANSMISSION IN HD UNITS

Health care procedures related to nosocomial infections in HD units represent the key factors in HCV transmission. In HD facilities, the most common lapses of healthcare quality are contamination of dialysis systems, inadequate disinfection and cleaning of environmental surfaces, improper contact of health care staff with equipment and patients, and mishandling of parenteral medications [16,17]. The guidelines for preventing HCV infection in HD settings recommend fundamental infection control practices and routine screening of patients undergoing HD for HCV. Isolating patients with HCV infection or those using dedicated machines for such patients are not advocated, except as necessary during local outbreaks [17,18].

NATURAL COURSE OF HCV INFECTION IN PATIENTS WITH ESRD

In patients with ESRD, chronic HCV infection usually takes an insidious clinical course. Early diagnosis and identification of individuals at greater risk for fibrosis progression has been suggested [19]. This is important because during the interval between diagnosis and transplantation there is increased mortality risk because of higher incidence of hepatic and non-hepatic complications. Thus, patients with HCV infection and with ESRD should be followed up to determine viral load, do HCV genotyping, assess the extent of hepatic fibrosis, and establish optimal treatment strategies. Treating HCV infection before KT helps to prevent posttransplantation complications and reduce mortality [20–22].

TREATMENT OF HCV IN PATIENTS WITH ESRD

The primary goal of HCV treatment in patients with ESRD is to achieve sustained viral response (SVR) before KT. Another important reason for treatment is to attain SVR before KT because of the concern that antiviral therapy administered posttransplantation is associated with a high risk of graft rejection [23,24]. Other goal is to reduce the likelihood of HCV-related complications in the liver and other organs/systems [23,24].

Monotherapy With Standard Interferon or Pegylated Interferon

Three meta-analyses have indicated that SVR, side effects, and withdrawal rates in patients with ESRD vary according to treatment with interferon (IFN) and pegylated interferon (peg-IFN). A meta-analysis [11] evaluating results from 645 patients showed that the overall SVR rate was 40%; in the subset with HCV genotype 1, the SVR rate was 33% and dropout rates were 19% in the group that received IFN and 27% in the group that received peg-IFN. A typical flulike syndrome was the most common side effect. This occurred in 41% of patients and required withdrawal of antiviral treatment in 11%. Another meta-analysis by Gordon et al. [25] involved 546 patients on HD who had chronic HCV infection and were treated with either IFN or peg-IFN, with or without ribavirin. Only 49 individuals received peg-IFN and ribavirin. The overall SVR rates were 41% for the IFN group (95% CI: 33%–49%) and 37% for the peg-IFN group (95% CI: 9%–77%). The frequencies of treatment discontinuation were 26% for the IFN group (95% CI: 20%–34%) and 28% for the peg-IFN group (95% CI: 12%–53%). The main side effects were fatigue/weakness and loss of appetite. The authors also found that higher dose of IFN, lower HCV RNA load before treatment, early stage of hepatic fibrosis, and HCV genotype other than genotype 1 were associated with higher SVR rates. A third meta-analysis evaluated data from 770 patients on HD with chronic HCV infection, 491 of whom received IFN α-2a or IFN α-2b and 279 of whom received peg-IFN α-2a or peg-IFN α-2b. The corresponding SVR rates for these two groups were 39.1% (95% CI: 32.1%–46.1%) and 39.3% (95% CI: 26.5%–52.1%), and the corresponding dropout rates were 22.6% (95% CI: 10.4%–34.8%) and 29.7% (95% CI: 21.7%–37.7%). Age younger than 40 years was significantly associated with SVR (OR = 2.17; 95% CI: 1.03–4.50) [26].

IFN or peg-IFN With Ribavirin

The combination of peg-IFN and ribavirin is considered the gold standard therapy for patients with chronic HCV who have normal renal function; however, several physicians are reluctant to use ribavirin in patients with ESRD or in those who are on HD because of fear of hemolytic anemia, which can be exacerbated in the presence of kidney dysfunction. Because ribavirin is filtered by the kidneys, its clearance is impaired in patients with ESRD, and this agent is not removed by dialysis [27]. Despite the fact that ribavirin is contraindicated in the setting of renal failure, this drug can be used at markedly reduced daily doses and with careful monitoring for anemia. Patients can be started on a low dose of ribavirin, and doses can be increased gradually as long as side effects are manageable [27,28]. The largest series to date on the combined use of peg-IFN α-2a (135 µg/wk) plus ribavirin (200 mg daily to every other day) for 48 wk in 35 patients on HD demonstrated an SVR rate of 97% (34 of the 35 patients) and a dropout rate of 14%. Only one patient developed severe anemia and had to be weaned off. The dose of ribavirin should be adjusted based on the target plasma level, which has been identified as 10–15 mcmol/L in patients with normal kidney function. For patients with ESRD, the average dose of ribavirin can be 200 mg daily, but some individuals can only tolerate 200 mg every other day. Assays for monitoring plasma ribavirin levels are not routinely available. Even when therapeutic ribavirin levels are maintained, the potential for anemia in patients on HD cannot be ruled out. Recombinant human erythropoietin or blood transfusions can be varied while maintaining the desired ribavirin dosage, for these patients, as these measures can correct anemia and improve the quality of life during treatment [27].

Direct-Acting Antiviral Agents

Few trials showed that direct-acting antiviral agents (DAA) have yielded promising results in patients with ESRD. According to the 2016 European Association of Liver guidelines, HCV treatment should be provided according to the following recommendations [29]:

1. Patients with mild to moderate renal impairment

    Patients with mild to moderate renal impairment [estimated glomerular filtration rate (eGFR) $\geq$30 mL/min/1.73 m^2] with HCV infection. No dose adjustments of HCV DAAs are needed, but these patients should be carefully monitored.

2. Patients with severe renal impairment

    Patients with severe renal impairment (eGFR <30 mL/min/1.73 m^2) and patients with ESRD on HD should be carefully monitored by a multidisciplinary team. Sofosbuvir should be used with caution in patients with an eGFR <30 mL/min/1.73 m^2 or with ESRD because no dose recommendation can currently be given for these patients. Treatment of HCV in this group is designed according to HCV genotype.

    a. Patients with severe renal impairment (eGFR <30 mL/min/1.73 m^2) or with ESRD on HD without an indication for KT infected with HCV genotype 1a should be treated with the combination of ritonavir-boosted paritaprevir, ombitasvir, and dasabuvir for 12 weeks or with the combination of grazoprevir and elbasvir for 12 weeks, with daily ribavirin (200 mg/day) if the hemoglobin level is >10 g/dL at baseline.

    b. Patients with severe renal impairment (eGFR <30 mL/min/1.73 m^2) or with ESRD on HD without an indication for KT infected with HCV genotype 1b should be treated with the combination of ritonavir-boosted paritaprevir, ombitasvir, and dasabuvir for 12 weeks or with the combination of grazoprevir and elbasvir for 12 weeks, without ribavirin.

    c. Patients with severe renal impairment (eGFR <30 mL/min/1.73 m^2) or with ESRD on HD without an indication for KT infected with HCV genotype 4 should be treated with a combination of ritonavir-boosted paritaprevir and ombitasvir for 12 weeks with daily ribavirin (200 mg/day) if the hemoglobin level is >10 g/dL at baseline or with the combination of grazoprevir and elbasvir for 12 weeks without ribavirin.

    d. If treatment is urgently needed in patients with severe renal impairment (eGFR <30 mL/min/1.73 m^2) or with ESRD on HD without an indication for KT infected with HCV genotype 2, these patients should receive the fixed-dose combination of sofosbuvir and velpatasvir, or the combination of sofosbuvir and daclatasvir for 12 weeks without ribavirin. Renal function may worsen and should be carefully monitored and treatment should be interrupted immediately in case of deterioration.

    e. If treatment is urgently needed in patients with severe renal impairment (eGFR <30 mL/min/1.73 m^2) or with ESRD on HD without an indication for KT infected with HCV genotype 3, these patients should receive the fixed-dose combination of sofosbuvir and velpatasvir, or the combination of sofosbuvir and daclatasvir for 12 weeks with daily ribavirin (200 mg/day) if the hemoglobin level is >10 g/dL at baseline, or for 24 weeks without ribavirin. Renal function may worsen and should be carefully monitored and treatment should be interrupted immediately in case of deterioration.

In patients receiving ribavirin, hemoglobin levels should be carefully and frequently monitored and ribavirin administration should be interrupted in case of severe anemia (hemoglobin <8.5 g/dL). The use of erythropoietin and, eventually, blood transfusion, may be useful in patients with severe ribavirin-induced anemia. Patients with cirrhosis and those with a contraindication or who do not tolerate ribavirin may benefit from 24 weeks of these therapies without ribavirin. The risks versus benefits of treating patients with ESRD and an indication for KT before or after renal transplantation require individual assessment.

In conclusion, IFN-free DAAs play an important role in the eradication of hepatitis C among patients with kidney disease. However, DAAs are very expensive; in an era of cost containment, this is a crucial point in developing countries. Adverse drug reactions resulting from concomitantly administered medications are another ongoing concern for patients undergoing HCV treatment, particularly for patients with chronic kidney disease who have a heavy burden of comorbidities.

ABBREVIATIONS

DAAs Direct anti-viral agents
ESRD End stage renal disease
HD Hemodialysis
IFN Interferon
KT Kidney transplantation
PD Peritoneal dialysis
PEG-IFN Pegylated interferon
RRT Renal replacement therapy
SVR Sustained virologic response

REFERENCES

[1] Grassmann A, Gioberge S, Moeller S, Brown G. ESRD patients in 2004: global overview of patient numbers, treatment modalities and associated trends. Nephrol Dial Transpl 2005; 20:2587–93.

[2] Bambgoye E. Haemodialysis: management problems in developing countries, with Nigeria as a surrogate. Kidney Int 2003;63(Suppl. 83):S93–5.

[3] Arije A, Kadiri S, Akinkugbe OO. The viability of hemodialysis as a treatment option for renal failure in a developing economy. Afr J Med Sci 2000;29:311–4.

[4] Bello AK, Levin A, Tonelli M, Okpechi IG, Feehally J, Harris D, Jindal K, Salako BL, Rateb A, Osman MA, Qarni B, Saad S, Lunney M, Wiebe N, Ye F, Johnson DW. Assessment of global kidney health care Status. JAMA May 9, 2017;317(18):1864–81.

[5] Bamgboye EL. The challenges of ESRD care in developing economies: sub-Saharan African opportunities for significant improvement. Clin Nephrol 2016;13:18–22. 86(2016 Suppl. 1).

[6] Prasad N, Jha V. Hemodialysis in Asia. Kidney Dis (Basel) 2015;1(3):165–77.

[7] Mendizabal M, Reddy KR. Chronic hepatitis C and chronic kidney disease: advances, limitations and unchartered territories. J Viral Hepat 2017;24(6):442–53.

[8] Fissell RB, Bragg-Gresham JL, Woods JD, Jadoul M, Gillespie B, Hedderwick SA, Rayner HC, Greenwood RN, Akiba T, Young EW. Patterns of hepatitis C prevalence and seroconversion in hemodialysis units from three continents: the DOPPS. Kidney Int 2004;65:2335–42.

[9] Di Napoli A, Pezzotti P, Di Lallo D, Petrosillo N, Trivelloni C, Di Giulio S. Epidemiology of hepatitis C virus among long-term dialysis patients: a 9-year study in an Italian region. Am J Kidney Dis 2006;48:629–37.

[10] Sun J, Yu R, Zhu B, Wu J, Larsen S, Zhao W. Hepatitis C infection and related factors in hemodialysis patients in China: systematic review and meta-analysis. Ren Fail 2009; 31:610–20.

[11] Fabrizi F, Martin P, Dixit V, Bunnapradist S, Dulai G. Meta-analysis: effect of hepatitis C virus infection on mortality in dialysis. Aliment Pharmacol Ther 2004;20:1271–7.

[12] Sezer S, Ozdemir FN, Akcay A, Arat Z, Boyacioglu S, Haberal M. Renal transplantation offers a better survival in HCV-infected ESRD patients. Clin Transpl 2004;18:619–23.

[13] Ruhı Ç, Süleymanlar İ KH, Yilmaz VT, Çolak D, Dınçkan A, Gürkan A, Ersoy F, Yakupoğlu G, Süleymanlar G. The impact of hepatitis C virus infection on long-term outcome in renal transplant patients. Turk J Gastroenterol 2011;22:165–70.

[14] Roth D, Gaynor JJ, Reddy KR, Ciancio G, Sageshima J, Kupin W, Guerra G, Chen L, Burke GW. Effect of kidney transplantation on outcomes among patients with hepatitis C. J Am Soc Nephrol 2011;22:1152–60.

[15] Ingsathit A, Kamanamool N, Thakkinstian A, Sumethkul V. Survival advantage of kidney transplantation over dialysis in patients with hepatitis C: a systematic review and meta-analysis. Transplantation 2013;95:943–8.

[16] Recommendations for preventing transmission of infections among chronic hemodialysis patients. MMWR Recomm Rep 2001;50:1–43.

[17] Bianco A, Bova F, Nobile CG, Pileggi C, Pavia M. Healthcare workers and prevention of hepatitis C virus transmission: exploring knowledge, attitudes and evidence-based practices in hemodialysis units in Italy. BMC Infect Dis 2013;13:76.

[18] Tang S, Lai KN. Chronic viral hepatitis in hemodialysis patients. Hemodial Int 2005; 9:169–79.

[19] Kellner P, Anadol E, Hüneburg R, Hundt F, Bös D, Klein B, Woitas RP, Spengler U, Sauerbruch T, Trebicka J. The effect of hemodialysis on liver stiffness measurement: a single-center series. Eur J Gastroenterol Hepatol 2013;25:368–72.

[20] Botero RC. Should patients with chronic hepatitis C infection be transplanted? Transpl Proc 2004;36:1449–54.

[21] Morales JM, Bloom R, Roth D. Kidney transplantation in the patient with hepatitis C virus infection. Contrib Nephrol 2012;176:77–86.

[22] Carbone M, Cockwell P, Neuberger J. Hepatitis C and kidney transplantation. Int J Nephrol 2011;2011:59329.

[23] Esforzado N, Campistol JM. Treatment of hepatitis C in dialysis patients. Contrib Nephrol 2012;176:54–65.

[24] Rostaing L, Weclawiak H, Izopet J, Kamar N. Treatment of hepatitis C virus infection after kidney transplantation. Contrib Nephrol 2012;176:87–96.

[25] Gordon CE, Uhlig K, Lau J, Schmid CH, Levey AS, Wong JB. Interferon treatment in hemodialysis patients with chronic hepatitis C virus infection: a systematic review of the literature and meta-analysis of treatment efficacy and harms. Am J Kidney Dis 2008;51:263–77.

[26] Alavian SM, Tabatabaei SV. Meta-analysis of factors associated with sustained viral response in patients on hemodialysis treated with standard or pegylated interferon for hepatitis C infection. Iran J Kidney Dis 2010;4:181–94.

[27] Carrion AF, Fabrizi F, Martin P. Should ribavirin be used to treat hepatitis C in dialysis patients? Semin Dial 2011;24:272–4.

[28] Liu CH, Huang CF, Liu CJ, Dai CY, Liang CC, Huang JF, Hung PH, Tsai HB, Tsai MK, Chen SI, et al. Pegylated interferon-α2a with or without low-dose ribavirin for treatment-naive patients with hepatitis C virus genotype 1 receiving hemodialysis: a randomized trial. Ann Intern Med 2013;159:729–38.

[29] EASL recommendations for treatment of hepatitis C. 2016. Internet available at: http://www.easl.eu/medias/cpg/HCV2016/Summary.pdf.

FURTHER READING

[1] Rendina M, Schena A, Castellaneta NM, Losito F, Amoruso AC, Stallone G, Schena FP, Di Leo A, Francavilla A. The treatment of chronic hepatitis C with peginterferon alfa-2a (40 kDa) plus ribavirin in haemodialysed patients awaiting renal transplant. J Hepatol 2007; 46:768–74.

Chapter 5

Hepatitis C Screening and Diagnosis in Developing Countries

Huda H. Gaafar

Prince Sattam College of Medicine, Kharj, Kingdom of Saudi Arabia

Chapter Outline

SCREENING OF HEPATITIS C VIRUS INFECTION

Hepatitis C virus (HCV) is a global health problem with an estimated 130–150 million people infected. In developing countries, HCV represents a huge burden because of the high expenses of diagnosis and treatment [1]. HCV screening is crucial for the early diagnosis and treatment of hepatitis C infection. By detecting HCV infection early, antiviral treatment can be offered earlier in the course of the disease which is more effective than starting at a later stage [2]. Diagnostic tests are complex and expensive and can only be performed at centralized laboratories, which are few in low-resource settings.

Early detection together with counseling and lifestyle modifications may reduce the risk of transmission of HCV infection to other people. Screening is prioritized for individuals having risk factors for exposure to HCV including:

- Recipient of blood or blood components (red cells, platelets, fresh-frozen plasma).
- Recipient of blood from a HCV-positive donor.

Hepatitis C in Developing Countries. https://doi.org/10.1016/B978-0-12-803233-6.00016-3

- Injection drug user (past or present).
- Persons with the following associated conditions;
  - Persons with HIV infection,
  - Persons with hemophilia,
  - Persons who have ever been on hemodialysis
  - Persons with unexplained abnormal aminotransferase levels.
  - Children born to HCV-infected mothers.
  - Healthcare workers after a needle stick injury or mucosal exposure to HCV-positive blood.
  - Current sexual partners of HCV-infected persons.

Methods of HCV Screening

Screening for hepatitis C is performed by measuring antibody to HCV (anti-HCV) in serum. A positive test indicates that a person was previously exposed to hepatitis C. The currently available screening test has a sensitivity of at least 97% and a specificity of 100% [2,3]. A sensitivity of 97% indicates that the screening test will detect at least 97% of individuals who have been exposed. A specificity of 100% indicates that 100% of individuals without hepatitis C had a negative screening test with no false-positive test results. The screening test could be modified to increase the sensitivity and reduce the specificity because false-positive test results can be detected easily by measuring HCV RNA in the serum [2].

Different screening strategies have been implemented in different regions, based on the local epidemiology. Groups at higher risk of HCV infection should be tested. In regions with high endemicity, systematic one-time testing is recommended through regional or national screening. Screening for HCV infection is based on the detection of anti-HCV antibodies. In several resource-limited countries, high rates of false-positive, anti-HCV, serologic tests have been reported for both for patients with HCV monoinfection and HIV/HCV coinfection. Few studies have reported quantitative HCV RNA results, making it difficult to determine the level of sensitivity future assays would require for detection of active HCV infection. However, serologic rapid diagnostics are less expensive than quantitative and qualitative HCV RNA testing. Therefore, serologic tests are often used to screen prospective blood donors, which can result in inappropriate exclusion of uninfected donors in a region that already has the lowest per-population rate of blood donation and the highest wastage of donated units because of transmittable infectious diseases [10,11].

In addition to enzyme immunoassays (EIAs), rapid diagnostic tests (RDTs) can be used to screen for anti-HCV antibodies. RDTs use various matrices, including serum, plasma, and also finger-stick capillary whole blood or, for some of them, oral (crevicular) fluid, facilitating screening without the need for venous puncture, tube centrifugation, freezing, and skilled labor. RDTs are simple to perform at room temperature without specific instrumentation or extensive training. Dried blood spots can also be used to collect whole-blood specimens to perform EIA detection of anti-HCV antibodies in a central laboratory [3–11]. Table 5.1 summarizes some studies that compared the results of anti-HCV antibody testing to HCV RNA polymerase chain reaction.

Advantages of HCV Screening

Screening is essential for early identification of HCV-infected individuals. Diagnosing chronic hepatitis C before the development of advanced fibrosis, cirrhosis, or hepatocellular carcinoma provides HCV eradication in 90%–100% of patients treated with direct antiviral agents [12,13]. Thus, detecting individuals with hepatitis C before they develop signs or symptoms of the disease can have an important impact on their subsequent clinical course. Sustained virologic clearance for more than 6 months after treatment of hepatitis C is also associated with a reduction in all-cause mortality.

Barriers to HCV Screening in Developing Countries

Barriers to screening for hepatitis C in developing countries include limited access to health care, inadequate health insurance coverage, individuals' decreasing recall of past risky behaviors, lack of knowledge of hepatitis C prevalence, natural history, and available tests and treatments for hepatitis C at the provider level [14,15]. Another barrier to HCV elimination is the subclinical nature of chronic HCV. Many patients are unaware of their infection, with large variations across the different regions/countries. Some developing countries do not implement screening programs for HCV. Lack of accurate HCV prevalence and incidence data hamper correct estimates of the magnitude of the HCV.

Diagnosis of Acute HCV

The diagnosis of acute and chronic HCV infection is established by the detection of HCV RNA by a sensitive molecular method. Anti-HCV antibodies are detectable by EIA in the vast majority of patients with HCV infection, but EIA results may be negative in early acute hepatitis C and in profoundly immunosuppressed patients. Following spontaneous or treatment-induced viral clearance, anti-HCV antibodies persist in the absence of HCV RNA but may decline and finally disappear in some individuals [11]. The diagnosis of acute hepatitis C can be confidently made only if seroconversion to anti-HCV antibodies can be documented, because there is no serologic marker, which proves that HCV infection is in the de novo acquired acute phase. Not all patients with acute hepatitis C will be anti-HCV positive at diagnosis. In these cases, acute hepatitis C can be suspected if the clinical signs and symptoms are compatible with acute hepatitis (alanine aminotransferase >10 times the upper limit of normal, and jaundice) in the absence of a history of chronic liver disease or other causes of acute hepatitis, and/or if a likely recent source of transmission is identifiable

TABLE 5.1 Studies Comparing the Prevalence of HCV Infection by Anti-HCV Antibody and Nucleic Acid Testing in Sub-Saharan Africa

References	Study Population	HCV Assays Performed	Anti-HCV Prevalence % (n)	NAT Detection in Anti-HCV Reactive Samples % (n)
WHO [14]	Republic of the Congo Population-based	Ortho HCV 3.0 (ELISA Diagnostics Systems)	5.6% (50/887)	62% (31/50)
		Monolisa anti-HCV Plus v 2 (Bio-Rad)		
		Inno-Lia HCV Immunoassay (Innogenetics)		
		In-house PCR		
Konerman and Lok [15]	Uganda Hospitalized patients	RSA (Cortez Diagnostics)	5% (19/380 by RSA) 6.8% (26/380 by ADVIA) 13% (48/380 reactive by one or both assays)	29% (14/48)
		Advia Centaur HCV EIA (Siemens Diagnostics)		
		Branched DNA, Versant 3.0 (Siemens Diagnostics)		
		Amplicor PCR (Roche)		
		RIBA 3.0 (Chiron)		
Abreha et al. [16]	Ethiopia Voluntary testing and counselling center	Murex anti-HCV v 4.0 (Abbott-Murex Diagnostics)	3.63% (71/1954) 6.8% (50/734 HIV+) 1.7% (21/1220 HIV−)	25.4% (18/71)
		Abbott Real-time PCR (Abbott Molecular)		
Zeba et al. [17]	Burkina Faso Pregnant women	Acon HCV rapid test strips	2.14% (13/607) 2.38% (HIV+) 1.75% (HIV−)	100% (13/13)
		In-house PCR		

TABLE 5.1 Studies Comparing the Prevalence of HCV Infection by Anti-HCV Antibody and Nucleic Acid Testing in Sub-Saharan Africa—cont'd

Forbi et al. [18]	Nigeria Population-based	Immuno-chromatographic rapid assay (Shantha Biotechnics)	14.1% (73/519)	75.3% (55/73)
		In-house PCR		
Agbaji et al. [19]	Nigeria HIV + patients	Dia Pro EIA (Diagnostic Bioprobe)	18.3% (262/1431)	30.2% (79/262)
		Cobas Amplicor HCV Monitor 2.0 (Roche)		
Cable et al. [20]	South Africa Blood donations	Prism chemiluminescent analyzer (Abbott Diagnostics)	0.12% (976 samples reactive by Prism)	1.5% (15/976) 29.4% (15/51)
		Murex EIA (Abbott Diagnostics)	51 samples reactive with Prism and Murex	
		In-house PCR		
Iles et al. [21]	Democratic Republic of Congo Male members of uniformed services	Ortho 3.0 Enhanced SAVe (Ortho Clinical Diagnostics)	13.7% (41/299)	26.8% (11/41)
		In-house PCR		
Mullis et al. [22]	Uganda Rakai cohort	Ortho v 3.0 HCV EIA (Ortho Diagnostics)	7.6% (76/1000) 6.2% (31/500 in HIV+) 9.0% (45/500 in HIV−)	0% (0/76)
		Abbott Real-time PCR		

EIA, enzyme immunoassay; *HCV*, hepatitis C virus; *PCR*, polymerase chain reaction; *RSA*, rapid strip assay.

[11,12]. In all cases, HCV RNA can be detected during the acute phase, although brief intervals of undetectable HCV RNA may occur. HCV reinfection has been described after spontaneous or treatment-induced HCV clearance, essentially in patients at high risk of infection. Reinfection is defined by the reappearance of HCV RNA at least 6 months after an SVR and the demonstration that infection is caused by a different HCV strain [11].

Diagnosis of Chronic HCV

The diagnosis of chronic hepatitis C is based on the detection of both anti-HCV antibodies and HCV RNA in the presence of biological or histologic signs of chronic hepatitis. In the case of a newly acquired HCV infection, spontaneous viral clearance is very rare beyond 6 months of infection [14], and hence, the diagnosis of chronic hepatitis C can be made when HCV viremia persists beyond 6 months.

HCV core antigen is a surrogate marker of HCV replication. Core antigen detection can be used instead of HCV RNA detection to diagnose acute or chronic HCV infection. HCV core antigen assays are less sensitive than HCV RNA assays (lower limit of detection equivalent to ~500–3000 HCV RNA IU/mL, depending on the HCV genotype) [13–15,16,17]. As a result, HCV core antigen becomes detectable in peripheral blood a few days after HCV RNA in patients with acute hepatitis C. In rare cases, core antigen is undetectable in the presence of HCV RNA.

Diagnostics in Developing Countries

Diagnostics in low-resource countries need to be affordable, accurate, user-friendly, robust, and give rapid point-of-care (POC) test results. The World Health Organization (WHO) outlined the **ASSURED** criteria for POC tests. An ideal POC for resource-limited environments should be affordable, sensitive, specific, user-friendly, rapid and robust, equipment-free, and deliverable to end users criteria. Tools that satisfy the ASSURED criteria primarily aim to provide same-day diagnosis and facilitate immediate decision-making [5]. Quick receipt of results increases the number of people who know their HIV status, assists in delivering timely ART (especially to women in labor), and minimizes the number of patients lost to long-term follow-up [15]. In addition, devices whose use requires minimal training and that have a high throughput can widen the pool of end users and alleviate congestion at HIV testing centers [8]. Finally, such a device should be operable in resource-limited environs, which include those with unreliable electricity, nonsterile conditions, and a lack of trained personnel to perform the duties typically reserved for nurses and health workers. Further examples of target specifications, listed in Table 5.1, will vary on the basis of the needs and setting in which the device will be used. For example, a device that detects a lower viral load may be less useful than a device that provides semiquantitative but clinically useful viral load ranges (e.g., levels denoting a low, medium, and high risk of virologic failure). Hence, consultation with clinicians, health workers, and other end users can result in well-defined performance specifications that complement the ASSURED criteria and suit the specific needs of health professionals and patients.

REFERENCES

[1] Alter MJ, Kuhnert WL, Finelli L. Guidelines for laboratory testing and result reporting of antibody to hepatitis C virus. Centers for Disease Control and Prevention. MMWR Recomm Rep 2003;52:1–16.

[2] Centers for Disease Control and Prevention (CDC). Testing for HCV infection: an update of guidance for clinicians and laboratorians. MMWR Morb Mortal Wkly Rep 2013;62:362–5.

[3] Cloherty G, Talal A, Coller K, et al. Role of serologic and molecular diagnostic assays in identification and management of hepatitis C virus infection. J Clin Microbiol 2016;54: 265–73.

[4] Dawson GJ. The potential role of HCV core antigen testing in diagnosing HCV infection. Antivir Ther 2012;17:1431–5.

[5] Freiman JM, Tran TM, Schumacher SG, et al. Hepatitis C core antigen testing for diagnosis of hepatitis C virus infection: a systematic review and meta-analysis. Ann Intern Med 2016;165:345–55.

[6] Gardner LI, Metsch LR, Anderson-Mahoney P, et al. Efficacy of a brief case management intervention to link recently diagnosed HIV-infected persons to care. AIDS 2005;19:423–31.

[7] Gretch DR. Diagnostic tests for hepatitis C. Hepatology 1997;26(3 Suppl. 1):43S–7S.

[8] Gretch DR. Use and interpretation of HCV diagnostic tests in the clinical setting. Clin Liver Dis 1997;1:543–57.

[9] Horstmann E, Brown J, Islam F, Buck J, Agins BD. Retaining HIV-infected patients in care: where are we? Where do we go from here? Clin Infect Dis 2010;50:752–61.

[10] Kamili S, Drobeniuc J, Araujo AC, Hayden TM. Laboratory diagnostics for hepatitis C virus infection. Clin Infect Dis 2012;55(Suppl. 1):S43–8.

[11] Majid AM, Gretch DR. Current and future hepatitis C virus diagnostic testing: problems and advancements. Microbes Infect 2002;4:1227–36.

[12] Mayer KH. Introduction: linkage, engagement, and retention in HIV care—essential for optimal individual- and community-level outcomes in the era of highly active antiretroviral therapy. Clin Infect Dis 2011;52(Suppl. 2):S205–7.

[13] Pawlotsky JM. Use and interpretation of hepatitis C virus diagnostic assays. Clin Liver Dis 2003;7:127–37.

[14] WHO. Global hepatitis report; 2017. (Internet) Available at: www.who.int/hepatitis/publications/global-hepatitis-report2017/en/.

[15] Konerman MA, Lok A, Hepatitis C. treatment and barriers to eradication. Clin Transl Gastroenterol 2016;7(9):e193.

[16] Abreha T, Woldeamanuel Y, Pietsch C, Maier M, Asrat D, Abebe A, Hailegiorgis B, Aseffa A, Liebert UG. Genotypes and viral load of hepatitis C virus among persons attending a voluntary counseling and testing center in Ethiopia. J Med Virol May 2011;83(5):776–82.

[17] Zeba MT, Karou SD, Sagna T, Djigma F, Bisseye C, Ouermi D, Pietra V, Pignatelli S, Gnoula C, Sia JD, Moret R, Nikiema JB, Simpore J. HCV prevalence and co-infection with HIV among pregnant women in Saint Camille Medical Centre, Ouagadougou. Trop Med Int Health November 2011;16(11):1392–6.

[18] Forbi JC, Purdy MA, Campo DS, Vaughan G, Dimitrova ZE, Ganova-Raeva LM, Xia GL, Khudyakov YE. Epidemic history of hepatitis C virus infection in two remote communities in Nigeria, West Africa. J Gen Virol July 2012;93(Pt 7):1410–21.

[19] Agbaji O, Thio CL, Meloni S, Graham C, Muazu M, Nimzing L, Idoko J, Sankalé JL, Ekong E, Murphy R, Kanki P, Hawkins C. Impact of hepatitis C virus on HIV response to antiretroviral therapy in Nigeria. J Acquir Immune Defic Syndr February 1, 2013;62(2):204–7.

[20] Cable R, Lelie N, Bird A. Reduction of the risk of transfusion-transmitted viral infection by nucleic acid amplification testing in the Western Cape of South Africa: a 5-year review. Vox Sang February 2013;104(2):93–9.

[21] Iles JC, Abby Harrison GL, Lyons S, Djoko CF, Tamoufe U, Lebreton M, Schneider BS, Fair JN, Tshala FM, Kayembe PK, Muyembe JJ, Edidi-Basepeo S, Wolfe ND, Klenerman P, Simmonds P, Pybus OG. Hepatitis C virus infections in the Democratic Republic of Congo exhibit a cohort effect. Infect Genet Evol October 2013;19:386–94.

[22] Mullis CE, Laeyendecker O, Reynolds SJ, Ocama P, Quinn J, Boaz I, Gray RH, Kirk GD, Thomas DL, Quinn TC, Stabinski L. High frequency of false-positive hepatitis C virus enzyme-linked immunosorbent assay in Rakai, Uganda. Clin Infect Dis December 2013;57(12):1747–50.

Evolution of Hepatitis C Treatment

Chapter 6

Hepatitis C Treatment in the Era of Direct-Acting Antiviral Agents: Challenges in Developing Countries

Sanaa M. Kamal

Ain Shams Faculty of Medicine, Cairo, Egypt

Chapter Outline

Hepatitis C in Developing Countries. https://doi.org/10.1016/B978-0-12-803233-6.00017-5

INTRODUCTION

Hepatitis C virus (HCV) is a major cause of liver cirrhosis, end-stage liver disease, and liver transplantation throughout the world [1]. Approximately 170–200 million people equating to 3% of the world's population are infected with HCV [2]. The prevalence of HCV varies in different geographic regions. The prevalence of HCV infection is greater in Africa and Asia, with infection rates exceeding 5% [3–5]. Egypt has the highest prevalence of hepatitis C in the world, with 15% of the population affected [6–8]. In the United States, nearly 2% of the population is infected [9,10]. In Europe, the prevalence ranges from 0.1% in northern European countries to 1% in Mediterranean countries [10,11]. The immigration crisis may increase HCV prevalence in Europe, given that immigrants originate from countries with high rates of HCV [12].

NATURAL HISTORY AND OUTCOME OF HCV INFECTION

Acute HCV infection is mostly asymptomatic and rarely recognized clinically. Spontaneous viral clearance occurs in approximately 25% of patients [13,14]. The striking feature of HCV infection is its tendency to persist and evolve to chronic hepatitis. In some patients, chronic HCV progresses to liver cirrhosis and hepatocellular carcinoma (HCC) [14,15].

The outcome of HCV infection depends largely on several host, viral, and environmental factors that likely determine the outcome of infection. Early on, HCV infection triggers viral-associated molecular pattern (VAMP) receptors resulting in induction of an antiviral state through several pathways including modifying and degrading viral RNA, altering cellular vesicle trafficking, or through other not yet identified mechanisms [15–17]. Clearance of HCV is associated with the development of robust and multispecific CD4$^+$ and CD8$^+$ T-cell responses in blood and liver that can be maintained for years after recovery from acute disease [18–20]. In contrast, individuals who progress to chronic infection fail to mount such a response or may have inadequate production of the cytokines essential to control viral replication. Incomplete control of viral replication by CD8$^+$ T cells in the absence of sufficient memory CD4$^+$T cells leads to viral persistence and emergence of cytotoxic T lymphocytes (CTL) escape mutants [21–24].

Acute resolving hepatitis has been shown to be associated with HCV homogeneity, whereas progressing hepatitis is correlated with genetic diversity, presumably reflecting greater immune pressure during acute spontaneous clearance [25]. Polymorphisms of genes involved in innate immunity and those in genes encoding cytokines and other immunologic mediators may explain spontaneous recovery from acute HCV and influence the strength and nature of immune defense. Genes encoding the inhibitory natural killer (NK) cell receptor *KIR2DL3* and its human leukocyte antigen C group 1 (*HLA-C1*) ligand influenced spontaneous resolution of HCV infection, suggesting that inhibitory NK cell interactions are critical for antiviral immunity [26,27].

To date, there are no reliable methods to predict who will resolve acute HCV spontaneously and who will develop chronic HCV. Similarly, no reliable indicators exist for distinguishing patients with chronic hepatitis C who may develop cirrhosis or HCC. Thus, early effective treatment of patients with HCV is necessary for the prevention of progression of liver disease to cirrhosis and HCC. In the absence of a vaccine against HCV, efficient treatment is important for prevention of transmission along with adoption of infection control measures.

EVOLUTION OF HCV THERAPY

The ultimate goal of hepatitis C treatment is to reduce the occurrence of end-stage liver disease and its complications, including decompensated cirrhosis, liver transplantation, and HCC. Treatment success is assessed by sustained virologic response (SVR), defined by the presence of undetectable HCV RNA in blood several months after completing a course of treatment [28].

For two decades, the standard of care (SOC) for hepatitis C infection was interferon based. Interferon-alpha (IFNα) has potent antiviral activity because of its ability to induce IFN-simulated genes that encode proteins, which inhibit various stages of viral replication [29]. Type I IFNs bind a unique, ubiquitous, heterodimeric receptor consisting of interferon-alpha receptor 1/2 (IFNAR1/IFNAR2), resulting in the activation of signaling pathways and induction of a large number of IFN-stimulated genes (ISGs). ISG-encoded proteins mediate the antiviral and other effects of IFNs [29]. IFNAR1 and IFNAR2 are associated with the Janus-activated kinases (JAKs) tyrosine kinase 2 and JAK1, respectively. Binding of type I IFNs to their heterodimeric receptors leads to activation of JAKs, which results in tyrosine phosphorylation of signal transducer and activator of transcription 2 (STAT2) and STAT1; STAT1/STAT2 migrates into the nucleus and associates with the IFN regulatory factor 9 (IRF9), to form the STAT1–STAT2–IRF9 complex. This complex then binds IFN-stimulated response elements inside DNA to initiate the transcription of hundreds of different ISGs [30,31]. IFN regulatory genes facilitate both clearance of virus from infected cells and protection of neighboring uninfected cells from incoming viral progeny. The antiviral-associated protein kinase R plays an important role in restricting HCV 1a replication through regulating the nuclear factor (NF)-κB pathway [32,33].

Initially, chronic hepatitis C was treated by conventional IFN mono-therapy which yielded very poor response rates. Addition of the guanosine analog, ribavirin (RBV), to conventional IFN was associated with a slight improvement in SVR, although the improvement was far from satisfactory particularly in HCV genotypes 1 and 4. Pegylation of the IFN molecule resulted in modification of the pharmacokinetic profile of IFNα-2. Both PEG-IFNα-2a and PEG-IFNα-2b have slower absorption, more reduced distribution, and lower elimination rate than the nonpegylated IFNα. The maintained concentrations of PEG-IFNα allowed longer periods of viral inhibition with once-a-week dosing. PEG-IFN/RBV therapy resulted in improved SVR, defined as undetectable HCV RNA 24 weeks after comple-tion of treatment. With PEG-IFNα-2/RBV combination, response rates in genotypes 2 and 3 range between 70% and 80%. However, SVR rates in chronic HCV genotypes 1 and 4 infection are suboptimal. Adverse events are common with IFN-based regimens and include fatigue, flulike symp-toms, anxiety, skin rash, and gastrointestinal symptoms such as nausea and diarrhea. Hemolytic anemia is frequent because of the use of RBV. Some patients treated with PEG-IFN/RBV may develop cardiac arrhythmias or severe neuropsychiatric adverse events, depression, and suicidal tendency. The various adverse effects (AEs), the long duration of therapy, and the need to inject IFN reduce compliance and treatment adherence. These fac-tors have driven the urgent need to develop new treatments that are safer and more effective (Fig. 6.1). The discovery of direct-acting antiviral agents heralded the dawn of a new era of HCV cure, which was a dream.

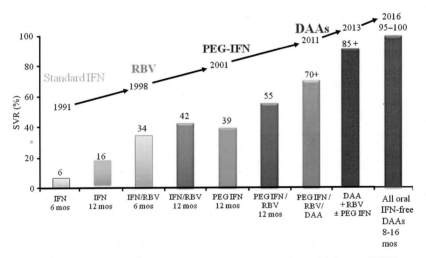

FIGURE 6.1   Evolution of chronic HCV therapies. *DAA*, direct-acting antiviral agent; *HCV*, hepa-titis C virus; *IFN*, interferon; *PEG*, pegylated; *RBV*, ribavirin; *SVR*, sustained virologic response.

DIRECT-ACTING ANTIVIRAL AGENTS

Direct-acting antiviral agents (DAAs) represent a revolution in HCV drug discovery. DAAs were developed to improve the SVR rates, reduce adverse events, and improve adherence to therapy among patients with HCV. DAAs were initially introduced as add-ons to the previous SOC consisting of PEF-IFNα/RBV. Recently, a breakthrough in HCV therapy has been achieved with the introduction of IFN-free all-oral DAAs, with SVR rates now in excess of 90% after 12 weeks of therapy for genotype 1 patients.

DAAs target specific steps within the HCV life cycle and disrupt viral replication in an attempt to terminate that cycle before its completion (Fig. 6.2) [34]. The first step in the life cycle of the virus is cell attachment and entry of HCV RNA through hepatocyte surface receptors. The HCV RNA is then translated to one polyprotein of 3010 amino acids that is subsequently cleaved by protease. It is then processed into four structural proteins (namely Core, E1, E2, and P7) and the nonstructural proteins [NS2-3 and NS3-4A proteases, NS3 helicase, and NS5B RNA-dependent RNA polymerase (RdRp)]. All these enzymes are essential for the replication of the virus and are potential drug discovery targets [35–37].

Goals of HCV and Endpoints of Treatment With DAAs

The goal of therapy is to eradicate HCV infection to prevent hepatic cirrhosis, decompensation of cirrhosis, HCC, and severe extrahepatic manifestations.

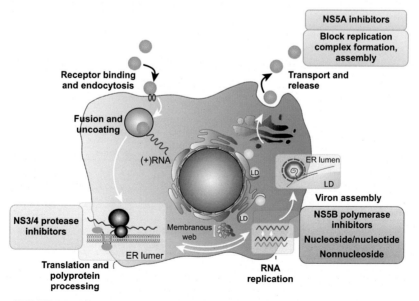

FIGURE 6.2   Hepatitis C virus replication cycle and targets of direct-acting antiviral agents.

The endpoint of therapy is undetectable HCV RNA in blood by a sensitive assay (with a lower limit of detection <15 IU/mL) at 12 weeks (SVR-12) and/or 24 weeks (SVR-24) after the end of treatment. Undetectable HCV core antigen 12 or 24 weeks after the end of therapy can be an alternative to HCV RNA testing to assess the SVR-12 or the SVR-24, respectively [38]. In patients with advanced fibrosis and cirrhosis, HCV eradication reduces the rate of decompensation and will reduce, albeit not abolish the risk of HCC. In these patients surveillance for HCC should be continued.

Classes of DAAs

There are four classes of DAAs, which are defined by their mechanism of action and therapeutic target. DAAs include NS5B nucleotide inhibitors, NS5B nonnucleoside inhibitors (NNIs), NS5A replication complex inhibitors, and NS3/4A protease inhibitors (PIs) (Fig. 6.3).

NS3/4A Protease Inhibitors

PIs block the activity NS3/4A serine protease, an enzyme which inhibits TRIF (TIR domain-containing adaptor-inducing IFN-β)-mediated Toll-like receptor signaling and Cardif-mediated retinoic acid–inducible gene 1 signaling resulting in impaired induction of IFNs and blocking viral elimination [39,40]. PIs have been grouped according to their resistance profile into first- and second-generation agents and into separate waves according to dosing, safety, and tolerability characteristics.

First-Generation PIs

Telaprevir (TPV) and boceprevir (BOC) were the first DAAs used for the treatment of HCV and represented the first generation of PIs. TPV or BOC was used

Characteristic	Protease Inhibitor[*]	Protease Inhibitor[**]	NS5A Inhibitor	Nuc Polymerase Inhibitor	Non-Nuc Polymerase Inhibitor
Resistance profile	●	○	○	◐	●
Pangenotypic efficacy	●	○	○	◐	○
Antiviral potency	○	◐	◐	◐	○
Adverse events	●	◐	◐	○	○

◐ Good profile   ○ Average profile   ● Least favorable profile

[*]First generation. [**]Second generation.

FIGURE 6.3   Comparison of potency and resistance of direct-acting antiviral agents.

in combination with PEG-IFN/RBV for the treatment of genotype 1 [40,41]. Although TPV or BOC regimen enhanced SVR rates, the clinical efficacy of the triple regimen was limited by narrow genotype specificity, low barrier to resistance, and drug–drug interactions. This regimen also increased adverse events such as rash and moderate to severe anemia to an extent that might require the reduction of the RBV dose. Patient adherence to and tolerability of triple therapy with BOC or TPV is a challenging issue as the two DAAs should be given three times daily with food. Triple therapy was not very effective in patients who did not respond to PEG-IFN/RBV. From an economic perspective, the triple therapy dramatically increased the costs of HCV treatment which are originally prohibitive. Thus, the clinical importance of these agents diminished substantially with the discovery of subsequent-generation PIs.

Second-Wave, First-Generation PIs

The second wave of PIs for HCV includes agents such as Simeprevir (SIM), Asunaprevir, Danoprevir (DNV), Faldaprevir, and Vaniprevir [42,43]. SIM (*Olysio*) is an NS3/4A HCV PI. SIM is a macrocyclic compound that noncovalently binds to and inhibits the NS3/4A HCV protease, a protein that is responsible for cleaving and processing the HCV-encoded polyprotein, a critical step in HCV viral life cycle [42,43]. SIM shows enhanced binding affinity and specificity to NS3/4A when compared with the first-generation PIs, TPV and BOC.

The safety and efficacy of SIM/PEG-IFN/RBV combination was investigated in treatment-naive patients with HCV genotype 1 infection (PILLAR trial) [44]. Enrolled patients received different SIM doses administered oncedaily (QD) with PEF-IFNα-2a/RBV. According to response-guided therapy (RGT) criteria, 79.2%–86.1% of SIM-treated patients completed treatment by week 24; 85.2%–95.6% of these subsequently achieved SVR. The safety profile of triple therapy with SIM was found to be comparable with that of PEG-IFN/RBV combination therapy [44]. In the QUEST 1 and QUEST 2 studies [45,46], treatment-naive genotype 1 patients were randomized to receive either triple therapy with SIM plus PEG-IFN/RBV using a RGT approach or SOC (48 weeks of PEG-IFN/RBV with placebo control). SVR-12 rates were 81% in the SIM arm versus 50% in the control arm. The majority of SIM-treated patients met the RGT and received 24 weeks of treatment and 86% of these patients achieved an SVR-12.

SIM also enhanced SVR in treatment-experienced patients. In the ASPIRE trial [47], treatment-experienced patients (with prior failure to PEG/RBV) were randomized to receive placebo plus PEG/RBV or one of six regimens consisting of SIM plus PEG-IFNα-2a/RBV. In the SIM-treated patients, the SVR-24 rates ranged from 61% to 80% (according to the regimen used), which was significantly higher than the 23% SVR in patients treated with PEG-IFN/RBV. The safety profile observed among patients in the SIM arm was similar to the safety profile for patients in the placebo arm. These results were supported by those of another clinical trial conducted on treatment-experienced HCV genotype 1 patients with a history of viral relapse. The overall SVR-12 in patients treated with SIM/PEG-IFN/RBV was of 79% compared with 36% for the PEG-IFN/RBV arm. Patients with advanced fibrosis (F3–F4 by METAVIR) also had superior SVR-12 rates with the addition of SIM (73% SVR-12 compared with 24% in control arm) [48].

The efficacy of SIM/PEG-IFN/RBV in treatment-naive and treatment-experienced patients with chronic HCV genotype 4 was evaluated in the RESTORE trial. Overall, 65.4% of the patients achieved an SRV12. The SVR-12 rates varied by treatment group, being 83% in treatment-naive patients, 86% in treatment-experienced patients with prior relapse, 60% in prior partial responders, and 40% in prior null responders [49].

These trials showed that SIM-based PEG-IFN/RBV triple therapy was effective, well-tolerated, and safe. However, the fast-paced HCV drug discovery paved the way to new IFN-free combinations, which combine efficacy, safety, and convenience. Thus, SIM was included with other DAAs such as Sofosbuvir (SOF) to form one of the earliest IFN-free combinations (discussed later).

Danoprevir (DNV) is a highly selective and potent second-wave inhibitor of HCV NS3/4A protease. Coadministration of 100 mg of ritonavir with DNV has been shown to optimize the pharmacokinetics of DNV, allowing for lower dosing and better antiviral activity The DAUPHINE trial [50] evaluated three different dosages of DNV/r: 50, 100, and 200 mg DNV, boosted with 100 mg ritonavir, assumed twice a day for 24 weeks. A study arm also explored DNV/r 100/100 mg, in a RGT algorithm, in which patients reaching an RVR received a total of 12 weeks of treatment. Overall, better SVR rates were achieved in higher dosage arms compared with lower dosage arms. SVR rates decreased with decreasing dosage of DNV/r as follows: 89.1%, 78.5%, and 69.1% [50]. Faldaprevir was evaluated in IFN-free regimen in combination with Deleobuvir, an NS5B nonnucleoside polymerase inhibitor and RBV, in patients with HCV-1b. The combination was highly efficacious, with 95% achieving SVR-12 including patients with compensated cirrhosis [51].

Taken together, the second-wave, first-generation PIs offer several benefits over the first-generation PIs, TVR, and BOC in terms of less side effect profile and more convenient dosing. However, these preparations still have low genetic barrier to resistance particularly for HCV-1a.

Second-Generation PIs

Recent second-generations PIs such as the macrocyclic compound Grazoprevir (GZR) offer several benefits over earlier PIs, including fewer drug–drug interactions, improved dosing schedules, and less frequent and less severe side effects. GZR is distinct from earlier-generation PIs in potency against a broader array of HCV genotypes, and its activity against some of the major resistance-associated variants (R155K and D168Y) resulting from failure with first-generation

PIs. GZR is available in combination with the NS5A inhibitor elbasvir (EBR). EBR/GZR (Zepatier) is available as fixed-dose tablets (50 mg/100 mg) and is prescribed as one tablet orally once daily, with or without food. The treatment duration and whether to take with or without RBV are dependent on genotypes and other patient variables [52].

The C-EDGE treatment-naive trial assessed the safety and efficacy of the fixed-dose combination of EBR–GZR (50/100 mg) in patients with genotype 1, 4, or 6 hepatitis C infection, with or without compensated cirrhosis. The overall SVR-12 rate was 95%, with rates of 92% for genotype 1a, 99% for genotype 1b, 100% for genotype 4, and 80% for genotype 6. No statistically significant difference in SVR-12 was found between cirrhotic and noncirrhotic patients [53]. The C-EDGE CO-STAR trial enrolled treatment-naive patients who inject drugs and were infected with chronic HCV genotype 1, 4, or 6. In this difficult-to-treat cohort, 95% were SVR-12 [54].

Treatment-naive patients with compensated cirrhosis and treatment-experienced patients with a prior null response to PEG plus RBV were randomized to receive EBR plus GZR, with or without RBV, for 12 or 18 weeks. The SVR-12 rates ranged between 90% and 97% in cirrhotics and 94% in null responder cirrhotic patients. The SVR-12 was 100% for genotype 1b. No additional benefit was achieved by adding RBV to EBR plus GZR in a subset of patients [55]. Treatment-experienced patients with genotype 1 HCV with previous failure of PEG-IFN/RBV and an earlier-generation PI (BOC, TPV, or SIM) achieved SVR-24 of 96% when treated with EBR plus GZR and RBV [56].

In the C-EDGE coinfection trial, patients with chronic hepatitis C genotype 1, 4, or 6 and HIV coinfection received EBR–GZR once daily for 12 weeks. Patients were on antiretroviral therapy (ART) with HIV viral suppression and the median CD4 cell count was 568 cells/mm^3. The overall SVR-12 rate was 96%, with the breakdown by genotype SVR-12 rates being 96.5%, 95.5%, and 96.4% for genotypes 1a, 1b, and 4, respectively. All cirrhotic patients achieved an SVR-12 [57].

Thus, Zepatier is active against a broad array of HCV genotypes including genotypes 1, 4, and 6 and some of the major resistance-associated variants (R155K and D168Y) resulting from failure with first-generation PIs. EBR–GZR is generally well tolerated; however, the AEs reported include headache, nausea, fatigue, decreased appetite, anemia, pyrexia, and elevations of ALT.

NS5B Nucleoside Polymerase Inhibitors

NS5B is an RdRp involved in posttranslational processing, which is vital for HCV replication. Nucleoside polymerase inhibitors (NPIs) are analogs of natural substrates that bind the active site of NS5B and terminate viral RNA chain generation. Given that the structure of NS5B is highly conserved across all HCV genotypes, NPIs are effective against all genotypes. NPIs show high antiviral activities in all genotypes and provide a high genetic barrier to resistance. Thus, NPIs are included in several efficacious all-oral combination therapies.

Polymerase inhibitors are categorized according to their mode of action and specificity into NPIs and non-NPIs. These two classes generally differ in specificity. Nucleoside inhibitors (NIs) bind to the catalytic site of the RNA polymerase causing chain termination. NNIs bind to a less conserved site resulting in a conformational change that distorts the positioning of residues binding RNA resulting in inhibition of polymerization [58,59].

Sofosbuvir (SOVALDI)

Sofosbuvir (SOF) is a nucleoside analog inhibitor of HCV NS5B polymerase. The triphosphate form of SOF mimics the natural cellular uridine nucleotide and is incorporated by the HCV RNA polymerase into the elongating RNA primer strand, resulting in viral chain termination. SOF is a prodrug that is rapidly converted after oral intake to GS-331007, which is taken up by hepatocytes. The cellular kinases convert GS-331007 to its pharmacologically active uridine analog 5′-triphosphate form (GS-461203) that is incorporated by the HCV RNA polymerase into the elongating RNA primer strand, resulting in chain termination. SOF is a potent pangenotypic NS5B polymerase inhibitor with a high barrier to resistance. It is available as 400-mg tablets administered once a day with or without food. The discovery of SOF has been a breakthrough in HCV treatment. Currently, SOF represents the backbone in several IFN-free regimens for the treatment of chronic hepatitis caused by various HCV genotypes. Excretion of SOF is through the kidney (80%) [57,58,60].

Efficacy of SOF Plus PEG-IFN/RBV

The ATOMIC and ELECTRON (Arms 1–8) studies [61,62] established the effectiveness of a 12-week course of SOF plus PEG-IFN/RBV in treatment-naive patients with HCVgenotype-1 with SVR rates ranging between 87% and 100%. In genotypes 3, the SVR-12 rates were 71% with the 16-week SOF plus RBV regimen, 84% with 24 weeks of SOV plus RBV, and 93% with 12 weeks of SOF plus RBV plus PEG-IFN. For the patients with genotype 2 infections, the SVR-12 rates were 87% with the 16-week SOF plus RBV regimen, 100% with 24 weeks of SOF plus RBV, and 94% with 12 weeks of SOF plus RBV plus PEG-IFN [63].

Efficacy of IFN-Free SOF Regimen
SOF and SIM Combination (*Olysio*)

This combination is the earliest IFN-free regimen that reached optimal results in terms of SVR. Trials showed efficacy and safety of this drug combination in treatment-naive and treatment-experienced patients across several genotypes. In OPTIMIST-1 trial, SOF and SIM 8-week versus 12-week regimen were evaluated in treatment-naive and treatment-experienced patients with chronic HCV genotype. The overall SVR-12 rate was 97% and 83% in treatment-naive and treatment-experienced patients who were treated for 12 and 8 weeks, respectively [64]. These findings were further confirmed in the OPTIMIST-2 trial, which demonstrated that the 12-week regimen of SOF plus SIM is effective in treatment-naive and treatment-experienced patients with cirrhosis and HCV genotype 1 infection, with the exception that patients with genotype 1a and the baseline Q80K mutation have SVR rates of only 74% [65].

A large-cohort, prospective study on genotype 1 patients treated with SOF plus SIM for 12–16 week showed that the overall SVR rate was 84%. Model-adjusted estimates demonstrated that patients with cirrhosis, prior decompensation, and previous PI treatments were less likely to achieve an SVR. Addition of RBV enhances SVR rates in such patients [66].

Taken together, several clinical trials demonstrate that the all-oral 12-week regimen of SIM plus SOF is effective and well tolerated in treatment-naive and treatment-experienced HCV genotype 1 patient without and with cirrhosis. RBV may be needed in patients with decompensation and previous PI treatment failure. The frequent adverse events include fatigue, headache, nausea, rash, and insomnia. Serious adverse events and treatment discontinuation occur in only 3% of patients.

SOF With the NS5A Inhibitor Ledipasvir (Harvoni)

Treatment-Naive and Treatment-Experienced Genotype 1 Patients Gane et al. [67] evaluated an all-oral regimen comprising SOF with ledipasvir or the NS5B NNI GS-9669 in patients with genotype 1 HCV infection. SOF (400 mg once daily) and ledipasvir (90 mg once daily) plus RBV were given for 12 weeks to treatment-naive patients and prior null responders. SOF and GS-9669 (500 mg once daily) plus RBV were given for 12 weeks to treatment-naive patients and prior null responders. Additionally, prior null responders with cirrhosis were randomly assigned to groups given a fixed-dose combination of SOF and ledipasvir, with RBV or without RBV and a group of treatment-naive patients received SOF, ledipasvir, and RBV for 6 weeks. SVR-12 was 100%, 92%, and 68% in treatment-naive patients receiving SOF; those receiving SOV, ledipasvir, RBV, GS-9669, and RBV; and patients receiving 6 weeks of SOF, ledipasvir, and RBV respectively. All noncirrhotic prior null responders receiving 12 weeks of SOF along with another DAA plus RBV achieved SVR-12 of 100%.

In the NIAID SYNERGY (genotype 1) study [68], treatment-naive patients with genotype 1 chronic HCV were randomized to receive either ledipasvir–SOF for 12 weeks or ledipasvir–SOF (90–400 mg) plus the NNI NS5B inhibitor GS-9669 (500 mg once daily) for 6 weeks, or ledipasvir–SOF (90–400 mg) plus the NS3/4A PI GS-9451 (80 mg once daily) for 6 weeks. Patients in the 12-week ledipasvir–SOF arm with any stage of fibrosis could be enrolled in the study. The SVR-12 rates were 100%, 95%, and 95% in the ledipasvir–SOF arm, the ledipasvir–SOF plus GS-9669 group, and the ledipasvir–SOF plus GS-9451 group, respectively.

A trial [69] evaluated 8- and 12-week courses of the fixed-dose combination of ledipasvir (90 mg) and SOF (400 mg), with or without RBV in treatment-naive and treatment-experienced patients with chronic HCV genotype 1 infection. In all of the five study arms, SVR-12 was achieved in 95%–100% of patients. The regimen of ledipasvir–SOF was well tolerated; only one patient had a serious adverse event of anemia, thought to be related to RBV. A recent large study that enrolled 4365 genotype 1, treatment-naive, HCV-infected patients treated with LDV/SOF ± RBV demonstrated SVR rates of 91.3% and 92.0% (3191/3495) for LDV/SOF and LDV/SOF + RBV, respectively [70].

Thus, the combination of ledipasvir–SOF with or without RBV is highly effective in treatment-naive and treatment-experienced patients with chronic HCV genotype 1.

SOF–Ledipasvir in HCV Non–Genotype 1 Patients Patients with genotype 3 and 6 achieved good SVR rates when treated with ledipasvir–SOF. The SVR-12 responses in treatment-naive patients with genotype 3 were superior in the regimen with RBV (100%) than without RBV (64%). Among the treatment-experienced patients, 82% of patients treated with ledipasvir–SOF plus RBV achieved an SVR-12, and the SVR-12 rate was 96% in the patients with genotype 6 [71].

The NIAID SYNERGY (Genotype 4) trial enrolled treatment-naive and treatment-experienced patients with genotype 4 chronic HCV to receive ledipasvir–SOF for 12 weeks. Patients with compensated cirrhosis were allowed to enroll in the study. Overall, in the intent-to-treat, SVR was 95% [72]. A recent study that enrolled treatment-naive and treatment-experienced patients with chronic HCV genotype 4 revealed SVR-12 of 78% in patients treated with ledipasvir–SOF for 12 weeks and SVR-12 in patients treated with 24 weeks [73]. These findings suggest that further studies are still needed to optimize ledipasvir–SOF therapy in patients with different stages of chronic HCV genotype 4. To date, the efficacy and duration of DAAs in HCV genotype 4 has not been adequately studied and further trials are required to optimize therapy in this genotype.

A clinical trial assessed response to ledipasvir–SOF in 41 patients with chronic HCV genotype 5 (21 treatment-naive and 20 treatment-experienced). The overall SVR-12 was 95% in treatment-naive and treatment-experienced patients, while the SVR-12 was 97% in patients without cirrhosis and 89% in patients with cirrhosis [74].

HCV/HIV-Coinfected Patients The ERADICATE trial investigated the safety and efficacy of a 12-week regimen of ledipasvir–SOF in HCV treatment-naive patients with genotype 1 chronic hepatitis C who are coinfected with HIV. Patients on ART were allowed to receive tenofovir–emtricitabine plus either efavirenz, raltegravir, rilpivirine, and rilpivirine plus raltegravir or efavirenz plus raltegravir. In patients taking ART, SVR-12 was 97% [75]. The SVR-12 was 96.4% in

German HIV/HCV-coinfected patients [76]. A study investigated the efficacy and safety of ledipasvir–SOF plus RBV for 24 weeks in HCV/HIV-coinfected patients who relapsed after receiving 12 weeks of ledipasvir–SOF therapy. The SVR-12 was 89%, suggesting that ledipasvir–SOF can be an effective salvage therapy for patients for whom DAA treatment has failed [77].

Thus, ledipasvir–SOF is well tolerated and effective in patients with genotype 1 HCV and HIV coinfection. However, more studies are required to investigate the efficacy and safety of ledipasvir–SOF in the treatment of HIV patients infected with various HCV genotypes. Importantly, the drug–drug interactions between ledipasvir–SOF and ART need extensive investigations on large cohorts.

Retreatment of SOF Failures In the NIAID retreatment of SOF failures trial [78], patients with genotype 1 HCV who previously had failed a 24-week course of SOF plus RBV achieved SVR-12 ranging between 98% and 100% when retreated with a fixed-dose combination of ledipasvir–SOF for 12 weeks. Despite the small sample size in this study, the trial showed that the 12-week regimen of ledipasvir–SOF without or with RBV was well tolerated and shows promise as a treatment option for patients with prior SOF failure.

SOF–Velpatasvir (Epclusa)

SOF–Velpatasvir (VEL) (SOF 400 mg plus VEL 100 mg) is an oral, fixed-dose combination of SOF and the novel NS5A replication complex inhibitor VEL. VEL (formerly GS-5816) has potent in vitro anti-HCV activity across all genotypes at the picomolar level. The combination of SOF–VEL is the first, once-daily, single-tablet regimen with pangenotypic activity. SOF–VEL is indicated for patients with chronic HCV genotype 1 through 6 [79]. A clinical trial [80] assessed the efficacy and safety of the combination of the nucleotide polymerase inhibitor SOF, the NS5A inhibitor VEL, and the NS3/4A PI GS-9857 in patients with HCV genotype 1 infection. Among treatment-naive patients without cirrhosis, the SVR rates were 71% and 100% after 6 and 8 weeks of treatment, respectively. Among treatment-naive patients with cirrhosis, 94% achieved SVR-12 after 8 weeks therapy and 81% after 8 weeks of treatment with RBV. The SVR-12 rates were 100% in DAA-experienced patients without cirrhosis and with cirrhosis, respectively.

SOF–VEL showed high efficacy in non–genotype 1. ASTRAL-2 study demonstrated that the SVR-12 was 99% treatment-naive and treatment-experienced patients infected with HCV genotype 2 [81]. Among HCV with chronic HCV genotype 3 treated with SOF–VEL, the ASTRAL-3 trial showed that the SVR-12 rate were 93% for treatment-naive and 89% for treatment-experienced patients [82]. SVR-12 was 100% in patients with chronic HCV genotype 4 [83]. The ASTRAL-4 studies [81] demonstrated that SOF–VEL plus RBV was effective in achieving high SVR-12 rate in patients with decompensated cirrhosis [84].

The ASTRAL-5 study investigated the safety and efficacy of 12 weeks SOF–VEL in patients with HIV and HCV coinfection. Enrolled patients were infected with genotype 1, 2, 3, 4, or 6 HCV infection; 18% had compensated cirrhosis and 29% were treatment-experienced. The mean CD4 count was 583 cells/mm^3 and all patients had HIV viral suppression. The antiretroviral regimens included tenofovir disoproxil fumarate. The overall SVR-12 rate was 95%. The presence of cirrhosis or treatment experience did not negatively influence treatment response [85].

NONSTRUCTURAL PROTEIN 5A INHIBITORS

The nonstructural protein 5A (NS5A) protein is essential for replication and assembly of HCV. Inhibitors of NS5A block viral production at an early stage of assembly, so that no viral RNA or nucleocapsid protein is released [86]. Therefore, agents that block NS5B activity (polymerase inhibitors) inhibit the virus's RdRp [86]. NIs bind to RdRp's active site, whereas the NNIs bind to the enzyme outside the active site, inducing conformational changes that downregulate RdRp's activity. As a result of mechanistic and potency differences, the NIs tend to have broad potency against multiple HCV genotypes and are less likely to select for resistant strains than are the NNIs [87]. Cyclophilin is a host protein that interacts with NS5B and appears to promote the HCV protein's ability to bind.

Daclatasvir (Daklinza)

Daclatasvir HCV is a first-in-class inhibitor of the NS5A, a phosphoprotein that plays an important role in hepatitis C replication. The exact mechanism by which daclatasvir inhibits the NS5A replication complex is unclear, but it is believed that daclatasvir inhibits viral RNA replication and virion assembly. It may also inhibit phosphorylation of the NS4A, and therefore the formation and activation of the HCV replication complex. Based on in vitro data, daclatasvir has shown activity against HCV genotypes 1 through 6, with EC_{50} values ranging from picomolar to low nanomolar level against wild-type HCV [88].

When used in combination with SOF, with or without RBV, daclatasvir showed high efficacy in pangenotypic all-oral regimen. According to the results of the AI444040 and ALLY-3 trials [89,90], a 12-week regimen of daclatasvir plus SOF in patients with chronic HCV genotype 1 or 3 infection without cirrhosis resulted in high SVR-12 rates, regardless of prior treatment experience. The ALLY-3 [90] trial demonstrated high SVR-12 rates with a 12- or 16-week regimen of daclatasvir plus SOF and RBV in patients with chronic HCV genotype 3 infections and advanced fibrosis or compensated cirrhosis. A daclatasvir plus SOF–based regimen demonstrated efficacy in patients with chronic HCV genotype 1, 3, or 4 infection and advanced cirrhosis or posttransplant recurrence in the ALLY-1 trial [91], and in patients co-infected with HCV genotype 1, 3, or 4 and HIV-1 in the ALLY-2 trial [92].

Daclatasvir plus SOF with or without RBV was generally well tolerated. Fatigue, headache, nausea, and diarrhea were the adverse events reported in some patients treated with daclatasvir [91, 92, 93]. Daclatasvir and SOF combination can potentially cause serious bradycardia when coadministered with amiodarone. Given that daclatasvir is a substrate of CYP3A, it is contraindicated for use with drugs that are strong inducers of CYP3A, including phenytoin, carbamazepine, and rifampin [93].

Data from clinical trials showed resistance-associated substitutions (RASs) in the *NS5A* gene [94]. Thus, the American Association for the Study of Liver Diseases (AASLD)/Infectious Diseases Society of America (IDSA) HCV Guidance Panel recommends testing for these substitutions when NS5A inhibitors fail [95]. Baseline *NS5A* polymorphisms may also impact the emergence of NS5A resistance [95].

Taken together, daclatasvir plus SOF with or without RBV is an important option for use in treatment-naive or treatment-experienced patients with chronic HCV genotype 1, 3, or 4 infections, including patients with advanced liver disease, posttransplant recurrence, and HIV-1 coinfection. Daclatasvir with SOF is a particularly useful RBV-free oral option for genotype 3 patients. Before initiation of SOF plus daclatasvir therapy, testing for NS5A polymorphisms is recommended in patients with genotype 1a and cirrhosis.

Ledipasvir

Ledipasvir is a potent inhibitor of HCV NS5A, a viral phosphoprotein that plays a critical role in viral replication, assembly, and secretion [96]. The results of clinical trials assessing ledipasvir combinations with SOF have been discussed previously.

Coadministration of amiodarone and ledipasvir–SOF is not recommended given that severe cases of symptomatic bradycardia have been reported. Ledipasvir–SOF has significant drug–drug interactions with P-gp inducers such as rifampin, which may cause a significant reduction in the levels of ledipasvir and SOF and reduced efficacy of ledipasvir–SOF [96].

OMBITASVIR–PAITAPREVIR– RITONAVIR–DASABUVIR (VIEKIRA PAK)

The four medications ombitasvir, paritaprevir, ritonavir, and dasabuvir are combined as a fixed-dose tablet and the dasabuvir is a separate tablet. Ombitasvir is a NS5A inhibitor with potent pangenotypic picomolar antiviral activity, paritaprevir is an inhibitor of the NS3/4A serine protease, and dasabuvir is a nonnucleoside NS5B polymerase inhibitor. Ritonavir is a potent inhibitor of

CYP3A4 enzymes and is used as a pharmacologic booster for paritaprevir—it significantly increases peak and trough paritaprevir plasma concentrations and the area under the curve of paritaprevir [97].

PEARL III trial demonstrated SVR-12 rate of 99.5% in treatment-naive patients with chronic HCV genotype 1b treated with ombitasvir–paritaprevir–ritonavir and dasabuvir plus RBV and 99% in patients who received ombitasvir–paritaprevir–ritonavir and dasabuvir without RBV [98]. The TURQUOISE trial assessed the efficacy and safety of ombitasvir, paritaprevir, ritonavir, and dasabuvir plus RBV in HCV/HIV-1–coinfected patients for 12 or 24 weeks. The study enrolled HCV treatment-naive or PEG-IFN/RBV–experienced patients, with or without Child-Pugh A cirrhosis. Patients with CD4+ count ≥200 cells/ mm^3 or CD4+ percentage ≥14% and plasma HIV-1 RNA suppressed on a stable atazanavir- or raltegravir-inclusive antiretroviral regimen were included. Among patients treated with 3D + RBV for 12 weeks, 93.5% achieved SVR-12. Among patients receiving 24 weeks of treatment, 96.9% achieved EOTR (ombitasvir/paritaprevir/ritonavir + dasabuvir); the most common AEs were fatigue, insomnia, and nausea. Elevation in total bilirubin was the most common laboratory abnormality, occurring predominantly in patients receiving atazanavir.

This combination was effective in patients with liver transplant who have recurrent hepatitis C genotype 1 infection [99]. In patients with stage 4 or 5 renal disease and patients on dialysis treated with ombitasvir–paritaprevir–ritonavir and dasabuvir, EOT response was 100% and SVR-12 response was achieved in 85% of patients with genotype 1b [100].

Treatment of Different HCV Genotypes

According to the 2016 HCV treatment guidelines of the AASLD and the IDSA [95] and European Association of Study of Liver Diseases (EASL), [98] chronic HCV caused by any genotype can be efficiently treated using all-oral DAA IFN-free regimens.

HCV Genotype 1 (Fig. 6.4)

Optimizing the regimen of therapy for chronic HCV genotype 1 depends on several factors such as whether the patient is treatment naive or treatment experienced, the previous therapies provided, and the status of resistance in some cases. Given the high cost of IFN-free regimen and difficulties in access to this therapy in various countries, it is necessary to tailor therapy according to the patient population treated and the available therapies.

Ledipasvir–SOF Combination

- Treatment-naive patients with or without compensated cirrhosis are treated with a fixed-dose combination of SOF and ledipasvir for 12 weeks.
- Therapy may be shortened to 8 weeks in patients with a viral load <6 million IU/mL.

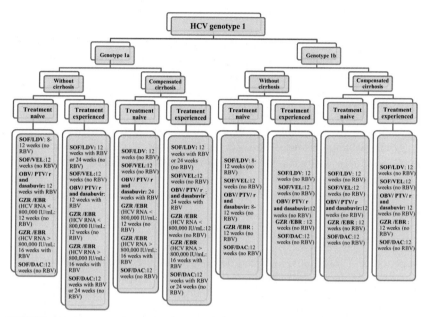

FIGURE 6.4   Treatment of HCV genotype 1. *DAA*, direct-acting antiviral agent; *DAC*, daclatasvir; *EBR*, elbasvir; *HCV*, hepatitis C virus; *GZR*, grazoprevir; *LDV*, ledipasvir; *OBV*, ombitasvir; *PTV*, paritaprevir; *r*, ritonavir; *SOF*, sofosbuvir; *VEL*, velpatasvir.

- Treatment-experienced, DAA-naive patients infected with genotype 1b with or without compensated cirrhosis are treated with the fixed-dose combination of SOF and ledipasvir for 12 weeks without RBV.
- Treatment-experienced, DAA-naive patients infected with genotype 1a with or without compensated cirrhosis are treated with the fixed-dose combination of SOF and ledipasvir for 12 weeks with daily, weight-based RBV (1000 or 1200 mg in patients <75 kg or =75 kg, respectively).
- If there is a contraindication to RBV, the treatment-experienced, DAA-naive patients are treated with Harvoni for 24 weeks.
- Treatment-experienced, DAA-naive patients infected with genotype 1a with or without compensated cirrhosis who have NS5A RASs and resistance to ledipasvir (M28A/G/T, Q30E/G/H/K/R, L31M/V, P32L/S, H58D, and/or Y93C/H/N/S) are treated with a fixed-dose combination of SOF and ledipasvir for 12 weeks with RBV.

SOF–VEL

- Genotype 1a, regardless of the presence of cirrhosis: SOF–VEL is prescribed for 12 weeks without RBV in treatment-naive or treatment-experienced patients.
- Genotype 1b, regardless of the presence of cirrhosis: SOF–VEL is prescribed for 12 weeks without RBV in treatment-naive or treatment-experienced patients.

EBR–GZR

EBR–GZR (50 mg/100 mg) therapy in chronic hepatitis C genotypes 1 is tailored according prior treatment experience and presence of baseline polymorphisms at amino acid positions 28, 30, 31, or 93.

- Genotype 1a, treatment-naive or PEG-IFN/RBV-experienced with no baseline NS5A polymorphisms: EBR–GZR is given for 12 weeks.
- Genotype 1a, treatment-naive or PEG-IFN/RBV-experienced with baseline NS5A polymorphisms: EBR–GZR plus RBV is prescribed for 16 weeks.
- Genotype 1b, treatment-naive or PEG-IFN/RBV-experienced: EBR–GZR is given for 12 weeks.
- Genotype 1a or 1b, PEG-IFN/RBV/PI-experienced: EBR–GZR plus RBV is given for 12 weeks.

SOF and Daclatasvir

In genotype 1 infections, SOF and daclatasvir are used with or without RBV depending on the patient population

- Genotype 1, without cirrhosis: daclatasvir plus SOF is prescribed for 12 weeks.
- Genotype 1 with compensated cirrhosis: daclatasvir plus SOF is given for 12 weeks.
- Genotype 1 with decompensated (Child-Pugh B or C) cirrhosis: daclatasvir plus SOF plus RBV is given for 12 weeks.

Ombitasvir–Paritaprevir–Ritonavir–Dasabuvir

In regions where this combination is available, the therapeutic strategy is recommended as follows:

- Genotype 1a, without cirrhosis: ombitasvir–paritaprevir–ritonavir and dasabuvir plus RBV is prescribed for 12 weeks.
- Genotype 1a, with cirrhosis: ombitasvir–paritaprevir–ritonavir and dasabuvir plus RBV for 24 weeks.
- Genotype 1b, without cirrhosis: ombitasvir–paritaprevir–ritonavir and dasabuvir for 12 weeks.
- Genotype 1b, with cirrhosis: ombitasvir–paritaprevir–ritonavir and dasabuvir plus RBV for 12 weeks.

Thus, Viekira Pak is prescribed with RBV except in patients without cirrhosis.

Genotype 2 (Fig. 6.5)

Chronic HCV genotype 2 treatment-naive or treatment-experienced patients are treated with either SOF–VEL for 12 weeks without RBV or SOF and daclatasvir without RBV.

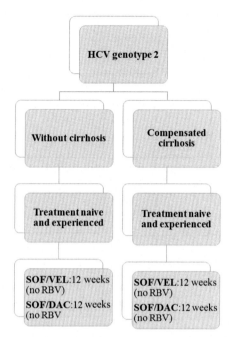

FIGURE 6.5 Treatment of HCV genotype 2. *DAC*, daclatasvir; *HCV*, hepatitis C virus; *SOF*, sofosbuvir; *VEL*, velpatasvir.

HCV Genotype 3 (Fig. 6.6)

Chronic HCV genotype 2 treatment naive patients are treated with either SOF–VEL for 12 weeks without RBV or SOF and daclatasvir without RBV. RBV is added for the therapy of treatment-experienced patients.

HCV Genotype 4 (Fig. 6.7)

Treatment-naive patients with chronic hepatitis C genotype 4 can be treated by one of the following regimens according to availability:

- SOF (400 mg)/ledipasvir (90 mg) for 12 weeks without RBV is prescribed for treatment-naive patients with or without compensated cirrhosis. In treatment-experienced patients, RBV is added as a daily weight-based dose (1000 or 1200 mg in patients <75 kg or =75 kg, respectively). SOF and ledipasvir for 24 weeks is recommended for treatment-experienced patients with or without compensated cirrhosis with contraindications to the use of RBV or with poor tolerance to RBV.
- SOF–VEL combination for 12 weeks without RBV is given to treatment-naive and treatment-experienced patients with chronic HCV genotype 4 with or without compensated cirrhosis.

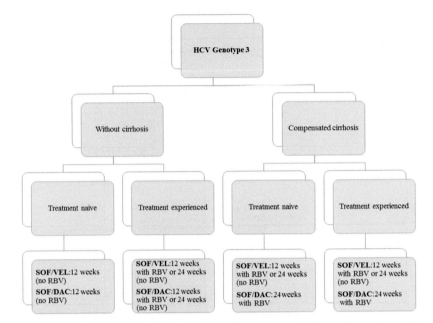

FIGURE 6.6 Treatment of HCV genotype 3. *DAC*, daclatasvir; *HCV*, hepatitis C virus; *SOF*, sofosbuvir; *VEL*, velpatasvir.

- Ombitasvir (12.5 mg), paritaprevir (75 mg), and ritonavir (50 mg) are given as two tablets once daily without dasabuvir to patients infected with HCV genotype 4 with and without compensated cirrhosis.
- GZR (100 mg) and EBR (50 mg) without RBV is prescribed as one tablet daily Treatment-naive patients infected with genotype 4 with or without compensated cirrhosis. Treatment-experienced patients infected with genotype 4 with or without compensated cirrhosis with an HCV RNA level at baseline >800,000 IU/mL are treated with GZR and EBR for 16 weeks with daily, weight-based RBV (1000 or 1200 mg in patients <75 kg or =75 kg, respectively).
- SOF (400 mg) and daclatasvir (60 mg) are given to treatment-naive patients with or without cirrhosis and should be treated with the combination of SOF and daclatasvir for 12 weeks without RBV. RBV (1000 or 1200 mg in patients <75 kg or =75 kg) is added for treatment-experienced patients with or without compensated cirrhosis.

HCV Genotype 5 or 6 (Fig. 6.8)

Treatment-naive patients with or without compensated cirrhosis. Patients with chronic HCV genotype 5 or 6 are treated with SOF and ledipasvir for 12 weeks without RBV. Treatment-experienced patients with or without compensated cirrhosis are treated with the combination of SOF and ledipasvir for 12 weeks with daily

Treatment of Chronic HCV Genotype 4

Treatment-naive genotype 4 without cirrhosis

Recommended regimen	Duration
Daily fixed-dose combination of glecaprevir (300 mg)/pibrentasvir (120 mg)	8 weeks
Daily fixed-dose combination of sofosbuvir (400 mg)/velpatasvir (100 mg)	12 weeks
Daily fixed-dose combination of elbasvir (50 mg)/grazoprevir (100 mg)	12 weeks
Daily fixed-dose combination of ledipasvir (90 mg)/sofosbuvir (400 mg)	12 weeks

Treatment-naive genotype 4 with Cirrhosis

Recommended regimen	Duration
Daily fixed-dose combination of sofosbuvir (400 mg)/velpatasvir (100 mg)	12 weeks
Daily fixed-dose combination of glecaprevir (300 mg)/pibrentasvir (120 mg)	12 weeks
Daily fixed-dose combination of elbasvir (50 mg)/grazoprevir (100 mg)	12 weeks
Daily fixed-dose combination of ledipasvir (90 mg)/sofosbuvir (400 mg)	12 weeks

Peginterferon/Ribavirin-experienced genotype 4 without Cirrhosis

Recommended regimen	Duration
Daily fixed-dose combination of sofosbuvir (400 mg)/velpatasvir (100 mg)	12 weeks
Daily fixed-dose combination of glecaprevir (300 mg)/pibrentasvir (120 mg)	8 weeks
Daily fixed-dose of elbasvir (50 mg)/grazoprevir (100 mg) for patients who experienced virologic relapse after prior peginterferon/ribavirin therapy	12 weeks

Peginterferon/Ribavirin-experienced genotype 4 with compensated cirrhosis

Recommended regimen	Duration
Daily fixed-dose combination of sofosbuvir (400 mg)/velpatasvir (100 mg)	12 weeks
Daily fixed-dose of elbasvir (50 mg)/grazoprevir (100 mg) for patients who experienced virologic relapse after prior peginterferon/ribavirin therapy	12 weeks

DAA-Experienced (Including NS5A Inhibitors), genotype 4 patients, With or Without compensated cirrhosis

Recommended regimen	Duration
Daily fixed-dose combination of sofosbuvir (400 mg)/velpatasvir (100mg)/voxilaprevir (100 mg)	12 weeks

FIGURE 6.7   Treatment of HCV genotype 4. *DAA*, direct-acting antiviral agent; *DAC*, daclatasvir; *EBR*, elbasvir; *HCV*, hepatitis C virus; *GZR*, grazoprevir; *LDV*, ledipasvir; *OBV*, ombitasvir; *PTV*, paritaprevir; *r*, ritonavir; *SOF*, sofosbuvir; *SIM*, simeprevir; *VEL*, velpatasvir.

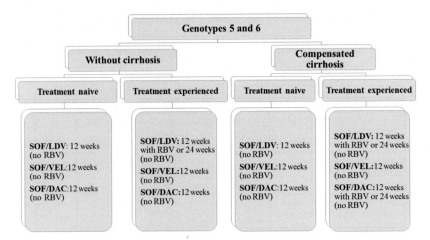

FIGURE 6.8 Treatment of HCV genotype 5,6. *DAC*, daclatasvir; *HCV*, hepatitis C virus; *SOF*, sofosbuvir; *VEL*, velpatasvir.

weight-based RBV (1000 or 1200 mg in patients <75 kg or=75 kg, respectively). Treatment-naive and treatment-experienced patients with or without compensated cirrhosis are treated with the fixed-dose combination of SOF and VEL for 12 weeks without RBV, and ribivarin is added in patients with treatment experienced patients.

PATIENTS WITH HCV AND HIV COINFECTION

Patients with HCV and HIV coinfection are treated according to genotype and prior treatment status as follows [95,98], and [99,101]:

1. Genotype 1a, treatment-naive patients may be treated with any of the following regimen:
   a. SOF–ledipasvir for 12 weeks without RBV
   b. SOF–VEL for 12 weeks without RBV
   c. Ombitasvir–paritaprevir–ritonavir and dasabuvir for 12 weeks with RBV
   d. GZR–EBR for 12 weeks without RBV if HCV RNA = 800,000 IU/mL or for 16 weeks with RBV if HCV RNA >800,000 IU/mL
   e. SOF–daclatasvir for 12 weeks without RBV
2. Genotype 1a, treatment-experienced patients may be treated with any of the following regimen:
   a. SOF–ledipasvir for 12 weeks with RBV or 24 weeks without RBV
   b. SOF–VEL for 12 weeks without RBV
   c. Ombitasvir–paritaprevir–ritonavir and dasabuvir for 12 weeks with RBV
   d. GZR–EBR for 12 weeks without RBV if HCV RNA = 800,000 IU/mL or for 16 weeks with RBV if HCV RNA >800,000 IU/mL
   e. SOF–daclatasvir for 12 weeks with RBV or 24 weeks without RBV

3. Genotype 1b, treatment-naive and treatment-experienced patients may be treated with any of the following regimens:
   a. SOF–ledipasvir for 12 weeks without RBV
   b. SOF–VEL for 12 weeks without RBV
   c. Ombitasvir/paritaprevir/ritonavir and dasabuvir for 12 weeks with RBV
   d. GZR/EBR for 12 weeks without RBV
   e. SOF–daclatasvir for 12 weeks without RBV
4. Genotype 2, treatment-naive and treatment-experienced patients may be treated with any of the following regimens:
   a. SOF–VEL for 12 weeks without RBV
   b. SOF–daclatasvir for 12 weeks without RBV
5. Genotype 3, treatment-naive and treatment-experienced patients may be treated with any of the following regimens:
   a. SOF–VEL for 12 weeks with RBV or 24 weeks without RBV
   b. SOF/daclatasvir for 12 weeks with RBV
6. Genotype 4 treatment-naive patients may be treated with any of the following regimens:
   a. SOF/ledipasvir for 12 weeks without RBV
   b. SOF–VEL for 12 weeks without RBV
   c. Ombitasvir/paritaprevir/ritonavir with RBV for 12 weeks
   d. GZR/EBR for 12 weeks without RBV
   e. SOF/daclatasvir for 12 weeks without RBV
   f. SOF and SIM for 12 weeks without RBV
7. Genotype 4 treatment-experienced patients may be treated with any of the following regimens:
   a. SOF/ledipasvir for 12 weeks with RBV and 24 weeks with RBV
   b. SOF–VEL for 12 weeks without RBV
   c. Ombitasvir/paritaprevir/ritonavir with RBV for 12 weeks
   d. GZR/EBR for 12 weeks without RBV if HCV RNA = 800,000 or for 16 weeks with RBV if HCV RNA >800,000 IU/mL
   e. SOF/daclatasvir for 12 weeks with RBV or 24 weeks without RBV
   f. SOF and SIM for 12 weeks with RBV or 24 weeks without RBV

Daclatasvir, dose requirement is needed with ritonavir-boosted atazanavir and efavirenz or etravirine. SIM should be used with antiretroviral drugs with which it does not have clinically significant interactions. In addition, daily fixed doses of combined SOF (400 mg)/VEL (100 mg) and of ofledipasvir (90 mg)/SOF (400 mg) are recommended. For combinations expected to increase tenofovir levels, baseline and ongoing assessment for tenofovir nephrotoxicity is recommended. Regarding HCV/HIV individuals, they should be treated and retreated as done for persons without HIV infection, after recognizing and managing interactions with antiretroviral medications [99,101].

TREATMENT OF PATIENTS WITH DECOMPENSATED CIRRHOSIS

Patients with decompensated cirrhosis and those awaiting liver transplantation are managed according to the HCV genotype. Patients with genotypes 1 and 4 are treated with daily fixed-dose combination of ledipasvir (90 mg)/SOF (400 mg) with low initial dose of RBV (600 mg, increased as tolerated) for 12 weeks. Another regimen is a daily fixed-dose combination of SOF (400 mg)/VEL (100 mg) with weight-based RBV for 12 weeks. Finally, daily doses of daclatasvir (60 mg) plus SOF (400 mg) with low initial dose of RBV (600 mg, increased as tolerated) for 12 weeks is recommended. For patients who are RBV-ineligible the recommended regimen is a daily fixed dose combination of SOF (400 mg)/VEL (100 mg) for 24 weeks. Another regimen is a combination of ledipasvir (90 mg)/SOF (400 mg) for 24 weeks. Patients who previously failed SOF-based treatment are given a combination of ledipasvir (90 mg)/SOF (400 mg) with low initial dose of RBV (600 mg, increased as tolerated) for 24 weeks [95,98,100]. Patients with HCV genotype 2 or 3 infection and decompensated cirrhosis are treated with daily fixed-dose combination SOF (400 mg)/VEL (100 mg) with weight-based RBV for 12 weeks [95,98,100].

PATIENTS WITH HCV RECURRENCE AFTER LIVER TRANSPLANTATION

Patients who develop HCV after transplantation and with compensated cirrhosis are treated with a daily fixed-dose combination of ledipasvir (90 mg)/SOF (400 mg) with weight-based RBV for 12 weeks. Treatment-naive patients with HCV genotype 1 or 4 infection in the allograft and with compensated liver disease and who are RBV ineligible can be treated by a daily fixed-dose combination of ledipasvir (90 mg)/SOF (400 mg) for 24 weeks [95,98,102]. Patients with HCV genotype 1 infection in the allograft, including those with compensated cirrhosis can receive daily SIM (150 mg) plus SOF (400 mg) with or without weight-based RBV for 12 weeks. For those with early-stage fibrosis, the recommended regimen is daily fixed-dose combination of paritaprevir (150 mg)/ritonavir (100 mg)/ombitasvir (25 mg) plus twice-daily dose of dasabuvir (250 mg) with weight-based RBV for 24 weeks. Treatment-naive and treatment-experienced patients with HCV genotype 2 infection in the allograft including those with compensated cirrhosis are treated with daclatasvir (60 mg) plus SOF (400 mg), with low initial dose of RBV (600 mg, increased as tolerated) for 12 weeks [91,95,98].

PATIENTS WITH HCV AND RENAL IMPAIRMENT

In patients with mild to moderate renal impairment, no dosage adjustment is required when using daclatasvir (60 mg), fixed-dose combination of ledipasvir (90 mg)/SOF (400 mg), fixed-dose combination of SOF (400 mg)/VEL (100 mg), or fixed-dose combination of paritaprevir (150 mg)/ritonavir (100 mg)/ombitasvir (25 mg) with (or without for HCV genotype 4 infection) twice-daily dose of dasabuvir (250 mg), SIM (150 mg), or SOF (400 mg) to treat or retreat HCV infection in patients with appropriate genotypes [95,98,103].

For patients with severe renal impairment or end stage-renal disease and patients with genotype 1a, 1b, or 4 infection and creatinine clearance (CrCl) below 30 mL/min, for whom treatment has been elected before kidney transplantation, daily fixed-dose combination of EBR (50 mg)/GZR (100 mg) for 12 weeks is recommended. For patients with genotype 1b infection with CrCl below 30 mL/min for whom the urgency to treat is high and treatment has been elected before kidney transplantation, daily fixed-dose combination of paritaprevir (150 mg)/ritonavir (100 mg)/ombitasvir (25 mg) plus twice-daily dose of dasabuvir (250 mg) for 12 weeks is recommended. For patients with HCV genotype 2, 3, 5, or 6 infection and CrCl below 30 mL/min for whom the urgency to treat is high and treatment has been elected before kidney transplantation, PEG-IFN and dose-adjusted RBV (200 mg daily) is recommended [103–106].

RETREATMENT OF PATIENTS WHO FAILED PREVIOUS THERAPY [56,95,98,107,108]

Patients who failed PEG-IFNα, RBV, and DAA or all DAA regimen are retreated according to the previous therapies and genotype as follows:

- Patients infected with HCV genotype 1 who failed after a triple combination regimen of PEG-IFNα, RBV and TPV, BOC, or SIM are treated with combination of SOF and ledipasvir or SOF and VEL, or SOF and daclatasvir, with RBV for 12 weeks.
- Patients who failed on SOF alone or SOF plus RBV or SOF plus PEG-IFNα/RBV can be retreated with any of the following:
  - Genotypes 1, 4, 5, or 6 can be treated with SOF and ledipasvir
  - All genotypes can be treated with SOF and VEL
  - Genotype 1 may be treated with ritonavir-boosted paritaprevir, ombitasvir, and dasabuvir
  - Genotype 4 is treated with ritonavir-boosted paritaprevir and ombitasvir, or SOF plus SIM
  - Genotypes 1 or 4 are treated with GZR and EBR for 24 weeks in F0–F2 patients with HCV RNA >800,000 IU/mL
  - All genotypes may be treated with SOF plus daclatasvir.

TREATMENT OF HCV AND HBV COINFECTION

The goal of therapy in hepatitis B virus (HBV) and HCV coinfection is to eradicate HCV infection and inhibit HBV replication. Evaluation of liver disease progression, predominance of one virus over another, and comorbidities are essential for optimal antiviral regimens. For patients with active hepatitis C, the same regimens following the same rules as for monoinfected patients should be applied based on AASLD and EASL recommendations [95,98]. For patients with active hepatitis B before, during, or after HCV clearance or with established cirrhosis, nucleoside or nucleotide analog (NA), tenofovir, or entecaviris is indicated [109,110]. Concurrent HBV nucleoside/NA therapy is indicated either if there is a potential risk of HBV reactivation during or after HCV clearance or if HBV replication is detectable at a significant level before the initiation of HCV treatment [111].

Patients should be carefully investigated for the replicative status of both HBV and HCV, and hepatitis delta virus infection before selecting the treatment strategy. When HCV is replicating and causes liver disease, it should be treated following the same rules as applied for HCV-monoinfected patients. There is a potential risk of HBV reactivation during or after HCV clearance. Before initiating DAA-based treatment for hepatitis C, patients should be tested for HBs antigen, anti-HBc antibodies, and anti-HBs antibodies. If HBs antigen is present or if HBV DNA is detectable in HBs antigen-negative, anti-HBc antibody-positive patients ("occult" hepatitis B), concurrent HBV nucleoside/NA therapy is indicated [95,98].

DAA RESISTANCE

Despite the great efficacy of the IFN-free DAA regimen, real-life experience revealed that approximately 5–10% of patients end up with virologic failure. Treatment failure raised the issue of resistance and occurrence of mutations. To date, the impact of such mutations on the treatment outcome is not clarified. It is not clear if the presence of mutations at baseline may independently lead to relapse [112]. HCV resistance–associated variants (RAVS) remains a challenging issue in HCV therapy. The prevalence of NS5A RAVs at baseline was shown to vary considerably across genotypes 1a, 1b, 3, and 4. Some studies showed that virologic failure tended to be more frequent when an NS5A Y93H substitution was present at baseline. RASs have been reported both in treatment-naive patients and following treatment with protease (NS3), phosphoprotein (NS5A), and polymerase (NS5B) inhibitors [112]. The different next-generation sequencing (NGS) technologies for HCV are critical for the identification of both viral genotype and resistance genetic motifs in the era of DAA therapies. A study [113] compared the ability of high-throughput NGS methods to generate full-length, deep, HCV sequence data sets and evaluated their utility for diagnostics and clinical assessment. The study showed that the consensus sequences generated by different NGS methods were generally concordant, and majority RAVs were consistently detected. However, methods differed in their ability to detect minor populations of RAVs. NGS provided a rapid, inexpensive method for generating whole HCV genomes to define infecting genotypes, RAVs, comprehensive viral strain analysis, and quasi-species diversity. Enrichment methods are particularly suited for high-throughput analysis while providing the genotype and information on potential DAA resistance [113].

Challenges to HCV Treatment in Developing Countries

The discovery of short-duration, safe, and highly effective regimens has opened up new horizons for HCV cure. However, real-life experience demonstrated some challenges such as emergence of mutations and management of special patient populations. Despite the optimism for the near future and the excellent efficacy, the prohibitive cost of such regimen is a great obstacle that interferes with accessibility of patients in countries with high HCV prevalence to the new IFN-free regimens. Thus, more efforts should be made to make IFN-free treatment cost-effective in all clinical scenarios and accessible to all patients.

Controlling the viral hepatitis epidemic and its consequences in developing countries is challenging but essential. Screening and access to antiviral therapy represent strong elements in the strategy for reduction of deaths from cirrhosis and HCC worldwide. Screening and assessing the severity of liver disease is crucial to select patients eligible for treatment. Moreover, in the absence of an HCV vaccine, availability of IFN-free regimens must be seriously considered for patients living in resource-limited settings. Global health and social justice should be taken into consideration when designing policies for the management of viral hepatitis and must also benefit patients living in resource-poor most endemic countries. Health policy-makers, medical doctors, scientists, and governments should share efforts to improve access to screening, care, and treatment for viral hepatitis. The implementation of clinical trials and national programs adapted to the local conditions of resource-poor settings, and the carrying out of cost–effectiveness studies are urgently needed to demonstrate the feasibility, safety, and benefits of such interventions in the future.

Prevention of Hepatitis C in Developing Countries

To date, no vaccines against HCV have been developed and immunoglobulin does not provide protection [114]. Several ongoing researches aim to develop safe and efficacious vaccines against HCV. An effective HCV vaccine would be based on two or several immunogens and might contain various epitopes [115]. Moreover, provocating intense and cross-reactive antiviral antibodies and multi-specific cellular immune responses are essential for an efficient HCV vaccine [116]. New HCV vaccine strategies, for instance, DNA and vector-based vaccines, peptides, and recombinant proteins are currently underway in phase I/II human clinical trials. Some of these vaccines provide an acceptable antiviral immunity in healthy volunteers and infected patients. Investigations for the improvement of DNA vaccines against viral and bacterial pathogens showed protection and prolonged immunity [117].

Recombinant Viral Vaccine Vectors

Most of the DNA-based vaccine research against various HCV proteins targeted either the humoral or cellular immune responses. However, more powerful vectors should be designed to generate both strong humoral and cellular immune responses against multiple epitopes within the structural and NS proteins. Recombinant viruses are efficient vehicle for DNA release that may cause a high level of recombinant protein expression in host cells [114–116]. Different recombinant viral vectors include adenovirus, vaccinia virus, and canarypox virus.

Peptide Vaccines

Class I major histocompatibility complex (MHC) molecules exist almost in all cell types and present only intracellularly generated peptide fragments to CD8+ T cytotoxic cells while class II MHC molecules exist on antigen-presenting cells and present antigenic peptides to T helper cells. Potential peptide vaccines would make use of small peptides present in the extracellular milieu and can bind directly to MHC class I or II molecules without undertaking the antigen processing ways. Accordingly, chemically synthesized peptides that are potent immunogenic antigens are being pursued as vaccine candidates for HCV [114,118].

Recombinant Protein Subunit Vaccines

A subunit vaccine containing recombinant HCV proteins can save from harm from infection or chronic infection by different HCV genotypes. The first effort to develop an HCV vaccine was directed toward generating a recombinant protein subunit vaccine. Because it has been shown for several flaviviruses that antibodies to the envelope protein can supply protection, recombinant HCV E1 and E2 proteins were used in early vaccination studies [119].

DNA Vaccine

One of the latest versions of vaccine is DNA-based immunization method [114,116–119]. DNA vaccines have shown superiority effects compared with conventional vaccines, such as recombinant protein-based vaccines and live weakened viruses. DNA immunization advantages include feasibility of production, DNA manipulating simplicity, and immune responses resulting primarily from different origins such as T helper cell and CTL, and antibody responses [120,121]. DNA vaccines are suitable for sequential vaccinations because their function is not influenced by preexisting antibody titers to the vector [114]. HCV is a very high variant virus that it is difficult to develop HCV vaccine [114,115]. NS3 gene is partially conserved and by inducing specific T-cell responses plays a major role in HCV clearance, making it a suitable candidate for T-cell–based vaccines because most researchers have concentrated on the specific CTL response stimulated by C and NS3 region proteins and protective antibodies induced by HCV E proteins. HCV core should be destroyed under the influence of immune response induced by a specific vaccine. It is a potent immunogen with anticore immune response that arises during the early stage of infection [122]. The HCV core protein might seem the obvious candidate for a therapeutic T-cell vaccine because this is the most highly conserved region of the translated HCV genome both within and between different HCV genotypes. However, studies have shown that the core protein can interfere with innate and adaptive anti-HCV immune responses. DNA-based vaccines are inferior to the traditional vaccines such as subunit vaccines because the intensity of the immune responses induced by DNA vaccines has been relatively weak; therefore, attempts are directed toward the development of new technique like co-delivery of novel cytokine IL-2, IL-7, IL-12, IL-15, and IL-18 adjuvants for circumventing this restriction [114,123,124].

HCV identification is indeed the most considerable recent development in viral disease. Because of the clinical significance of the disease, researches into the development of new therapeutic strategies are accentuated on the study of molecular properties of the virus. An efficient HCV vaccine should stimulate the different aspects of the immune response such as broad humoral, T helper, and CTL responses.

Because the HCV genome demonstrates high heterogeneity and mutagenicity, generating prophylactic or therapeutic vaccine for HCV is still an unsolved problem. Previous studies illustrated that the cellular immune responses might be essential for an efficient vaccine. New vaccine candidates, including DNA, peptide, recombinant protein, and vector-based vaccines, have been shown to have many advantages and lately have entered onto phase I/II human clinical trials. Some of these strategies provide an acceptable antiviral immunity in healthy volunteers and infected patients, but the challenge is to examining their effectiveness in infected or at-risk populations.

Primary prevention aims to reduce the risk for HCV transmission through the following procedures [115]:

- Avoiding unnecessary and unsafe injections, unsafe sharps waste collection, and disposal;
- Implementing strict infection control measures in health facilities;
- Screening of blood, plasma, organ, tissue, and semen donors;
- Controlling use of illicit drugs and preventing the sharing of injection equipment among drug users;
- Promoting protected sex with hepatitis C–infected subjects;
- Advising household contacts of HCV infects about the risk related to sharing of sharp personal items;
- Avoiding tattoos, piercings, and acupuncture performed with not sterilized equipment;
- Circumcision and other traditional health–related procedures should be performed by trained personnel applying universal infection control standards.
- Travelers to regions with high HCV endemicity should be advised against the use of medical, surgical, and dental equipment not adequately sterilized or disinfected and they should consider the risks about getting a tattoo or body piercing in areas where adequate sterilization or disinfection procedures might not be available or practiced.

Secondary prevention aims to reduce risks for chronic liver diseases in HCV-infected persons by identification, counseling, and testing and by providing appropriate medical management and antiviral therapy [77,81].

The determination of which risk groups to recommend for routine testing should be based on the knowledge of the epidemiologic link between a risk factor and acquiring HCV infection, prevalence of risk factors or behavior for infection in the population, and prevalence of infection among those with a risk factor or behavior [77].

Risk groups to test routinely should include injected illegal drugs users; subjects who received (1) clotting factor concentrates produced before the late

1980s, (2) blood from a donor who later tested positive for HCV infection, (3) transfusion of blood/blood components or an organ transplant before the early 1990s; and patients on long-term hemodialysis or with persistently abnormal alanine aminotransferase levels [77,82].

Based on recognized exposure, health-care, emergency, medical, and public safety workers who handle needle sticks or sharp instruments and those who have mucosal exposures to HCV-positive blood and children born to HCV-positive mothers should be tested routinely. Immunoglobulin and antiviral agents are not recommended for postexposure prophylaxis of hepatitis C.

In the United States, the Centers for Disease Control (CDC) strongly recommends that adults of all ages at risk for HCV infection should be tested, particularly those born during 1945–1965 without prior ascertainment of HCV risk should receive one-time testing for HCV [20].

Counseling directed to HCV-positive patients must include messages on how to protect the liver from further harm by receiving hepatitis A and B vaccination if susceptible, reducing alcohol consumption, and avoiding new medicines (including over-the-counter and herbal agents) without first checking with their health-care provider, and obtaining HIV risk testing [20]. Persons who are overweight/obese must lose weight and follow a healthy diet and a physically active lifestyle [20,83].

First of all, HCV-positive people must not donate blood, tissue, or semen and/or share devices that come into contact with blood, such as toothbrushes, dental appliances, razors, and nail clippers, to minimize the risk of transmission to others [20].

Epidemiologic surveillance represents the mainstay in the prevention of viral hepatitis because it provides the information for determining new infections and trends in incidence, changing patterns of transmission, and persons at highest risk for infection and for evaluating effectiveness or missed opportunities for prevention [77].

In the United States, each week, state and territorial health departments report cases of acute, symptomatic viral hepatitis to CDC [1].

Most of the European countries have a passive mandatory surveillance system for acute hepatitis having clinicians as the main source of data, using an electronic format at the national level, and collecting a similar set of basic data. Underreporting is common, but the exact extent is unknown [8].

Chronic hepatitis B and C, although accounting for the greatest burden of disease, are not reported by most states.

Serologic surveys carried out periodically can add information on geographical variations in prevalence of infections, populations at high risk, trends, and prevention programs [77].

ABBREVIATIONS

ART Antiretroviral
BOC Boceprevir
DAA Direct-acting antiviral
DNV Danoprevir

EBR-GZR Elbasvir-grazoprevir
EOTR Ombitasvir/paritaprevir/ritonavir + dasabuvir
GT Genotype
HCC Hepatocellular carcinoma
HCV Hepatitis C
HCV Hepatitis C virus
NNI Nonnucleotide polymerase inhibitor
NS5A Nonstructural protein 5A
Nuc Nucleotide polymerase inhibitor
OMV, RTV Ritonavir
PEG IFN Peginterferon
PI Protease inhibitor
RAV Resistance-associated variant.
RBV Ribavirin
RdRp RNA-dependent RNA polymerase
RGT Response-guided therapy
SIM Simeprevir
SOF Sofosbuvir
TPV Telaprevir
VEL Velpatasvir

REFERENCES

[1] World Health Organization. Hepatitis C. WHO fact sheet 164. Geneva (Switzerland): World Health Organization; 2000. Available from: http://www.who.int/mediacentre/factsheets/fs164/en/.

[2] Lavanchy D. The global burden of hepatitis C. Liver Int 2009;29(Suppl. 1):74–81.

[3] Shepard CW, Finelli L, Alter MJ. Global epidemiology of hepatitis C virus infection. Lancet Infect Dis 2005;5:558–67.

[4] Gower E, Estes C, Blach S, Razavi-Shearer K, Razavi H. Global epidemiology and genotype distribution of the hepatitis C virus infection. J Hepatol 2014;61(Suppl. 1):S45–57.

[5] Lavanchy D. Evolving epidemiology of hepatitis C virus. Clin Microbiol Infect 2011;17:107–15.

[6] Egyptian Ministry of Health. Annual Report 2007. Available from: http://www.mohp.gov.eg/Main.asp.

[7] Lehman EM, Wilson ML. Epidemic hepatitis C virus infection in Egypt: estimates of past incidence and future morbidity and mortality. J Viral Hepat 2009;16(9):650–8.

[8] Guerra J, Garenne M, Mohamed MK, Fontanet A. HCV burden of infection in Egypt: results from a nationwide survey. J Viral Hepat 2012;19:560–7.

[9] Armstrong GL, Wasley A, Simard EP, et al. The prevalence of hepatitis C virus infection in the United States, 1999 through 2002. Ann Intern Med 2006;144(10):705–14.

[10] Hanafiah KM, Groeger J, Flaxman AD, Wiersma ST. Global epidemiology of hepatitis C virus infection: new estimates of age-specific antibody to HCV seroprevalence. Hepatology 2013;57:1333–42.

[11] Desenclos JC. The challenge of hepatitis C surveillance in Europe. Euro Surveill 2003;8(5):99–110.

[12] Sharma S, Carballo M, Feld JJ, Janssen HLA. Immigration and viral hepatitis. J Hepatol 2015;63(2):515–22.

[13] Micallef JM, Kaldor JM, Dore GJ. Spontaneous viral clearance following acute hepatitis C infection: a systematic review of longitudinal studies. J Viral Hepat 2006;13(1):34–41.

[14] Hoofnagle JH. Course and outcome of hepatitis C. Hepatology 2002;36(5 Suppl. 1):S21–9.

[15] Seeff LB. The natural history of chronic hepatitis C virus infection. Clin Liver Dis 1997;1(3):587–602.

[16] Stone AE, Giugliano S, Schnell G, Cheng L, Leahy A, Golden-Mason L, Gale M, Rosen HR. Hepatitis C virus pathogen associated molecular pattern (PAMP) triggers production of lambda-interferons by human plasmacytoid dendritic cells. PLoS Pathol 2013;9(4):e1003316.

[17] Kell A, Stoddard M, Li H, Marcotrigiano J, Shaw JM, Gale Jr MM. Pathogen-associated molecular pattern recognition of hepatitis C virus transmitted/founder variants by RIG-I Is dependent on U-core length. J Virol 2015;89(21):11056–68.

[18] Thimme R, Oldach D, Chang KM, et al. Determinants of viral clearance and persistence during acute hepatitis C virus infection. J Exp Med 2001;194:1395–406.

[19] Kamal SM, Kassim SK, Ahmed AI, Mahmoud S, Bahnasy KA, Hafez TA, Aziz IA, Fathelbab IF, Mansour HM. Host and viral determinants of the outcome of exposure to HCV infection genotype 4: a large longitudinal study. Am J Gastroenterol February 2014;109(2):199–211.

[20] Keoshkerian E, Hunter M, Cameron B, Nguyen N, Sugden P, Bull R, Zekry A, Maher L, Seddiki N, Zaunders J, Kelleher A, Lloyd AR, HITS-p and HITS-c Investigators. Hepatitis C-specific effector and regulatory CD4 T-cell responses are associated with the outcomes of primary infection. J Viral Hepat August 25, 2016.

[21] Goh CC, Roggerson KM, Lee HC, Golden-Mason L, Rosen HR, Hahn YS. Hepatitis C virus-induced myeloid-derived suppressor cells suppress NK cell IFN-γ production by altering cellular metabolism via Arginase-1. J Immunol March 1, 2016;196(5):2283–92.

[22] Morishima C, Di Bisceglie AM, Rothman AL, Bonkovsky HL, Lindsay KL, Lee WM, Koziel MJ, Fontana RJ, Kim HY, Wright EC, HALT-C Trial Group. Antigen-specific T lymphocyte proliferation decreases over time in advanced chronic hepatitis C. J Viral Hepat June 2012;19(6):404–13.

[23] Kang W, Sung PS, Park SH, Yoon S, Chang DY, Kim S, Han KH, Kim JK, Rehermann B, Chwae YJ, Shin EC. Hepatitis C virus attenuates interferon-induced major histocompatibility complex class I expression and decreases CD8+ T cell effector functions. Gastroenterology May 2014;146(5):1351–60.

[24] Park SH, Rehermann B. Immune responses to HCV and other hepatitis viruses. Immunity January 16, 2014;40(1):13–24.

[25] Farci P. New insights into the HCV quasispecies and compartmentalization. Semin Liver Dis November 2011;31(4):356–74.

[26] Lunemann S, Martrus G, Hölzemer A, Chapel A, Ziegler M, Körner C, Garcia Beltran W, Carrington M, Wedemeyer H, Altfeld M. Sequence variations in HCV core-derived epitopes alter binding of KIR2DL3 to HLA-C*03:04 and modulate NK cell function. J Hepatol August 2016;65(2):252–8.

[27] Suppiah V, Gaudieri S, Armstrong NJ, O'Connor KS, Berg T, Weltman M, Abate ML, Spengler U, Bassendine M, Dore GJ, Irving WL, Powell E, Hellard M, Riordan S, Matthews G, Sheridan D, Nattermann J, Smedile A, Müller T, Hammond E, Dunn D, Negro F, Bochud PY, Mallal S, Ahlenstiel G, Stewart GJ, George J, Booth DR, International Hepatitis C Genetics Consortium (IHCGC). L28B, HLA-C, and KIR variants additively predict response to therapy in chronic hepatitis C virus infection in a European cohort: a cross-sectional study. PLoS Med September 2011;8(9):e1001092.

[28] Singal AG, Volk M, Jensen D, et al. A sustained viral response is associated with reduced liver related morbidity and mortality in patients with hepatitis C virus. Clin Gastroenterol Hepatol 2010;8:280–8.

[29] Takaoka A, Yanai H. Interferon signalling network in innate defence. Cell Microbiol 2006;8:907–22.

[30] Fensterl A, Sen GC. Interferons and viral infections. Biofactors 2009;35:14–20.

[31] deWeerd NA, Samarajiwa SA, Hertzog PJ. Type I interferon receptors: biochemistry and biological functions. J Biol Chem 2007;282:20053–7.

[32] Zhao W, Lee C, Piganis R, Plumlee C, de Weerd N, Hertzog PJ, et al. A conserved IFN-α receptor tyrosine motif directs the biological response to type I IFNs. J Immunol 2008;180:5483–9.

[33] Zhang L, Alter HJ, Wang H, Jia S, Wang E, Marincola FM, et al. The modulation of hepatitis C virus 1a replication by PKR is dependent on NF-κB mediated interferon β response in Huh7.5.1 cells. Virology 2013;438:28–36.

[34] Pokers P. New direct-acting antivirals in the development for hepatitis C virus infection. Therap Adv Gastroenterol May 2010;3(3):191–202.

[35] Kim CW, ChangK M. Hepatitis C virus: virology and life cycle. Clin Mol Hepatol March 2013;19(1):17–25.

[36] Bartenschlager R, Penin F, Lohmann V, Andre P. Assembly of infectious hepatitis C virus particles. Trends Microbiol 2011;19:95–103.

[37] Moradpour D, Penin F, Rice CM. Replication of hepatitis C virus. Nat Rev Microbiol 2007;5(6):453–63.

[38] Liang TJ, Ghany MG. Current and future therapies for hepatitis C virus infection. N Engl J Med 2013;368:1907–17.

[39] Rupp D, Bartenschlager R. Targets for antiviral therapy of hepatitis C. Semin Liver Dis 2014;34:9–21.

[40] Chang MH, Gordon LA, Fung HB. Boceprevir: a protease inhibitor for the treatment of hepatitis C. Clin Ther October 2012;34(10):2021–38.

[41] Matthews SJ, Lancaster JW. Telaprevir: a hepatitis C NS3/4A protease inhibitor. Clin Ther September 2012;34(9):1857–82.

[42] Asselah T, Marcellin P. Second-wave IFN-based triple therapy for HCV genotype 1 infection: simeprevir, faldaprevir and sofosbuvir. Liver Int 2014;34(Suppl. 1):60–8.

[43] Clark VC, Peter JA, Nelson DR. New therapeutic strategies in HCV: second-generation protease inhibitors. Liver Int 2013;33(Suppl. 1):80–4.

[44] Fried MW, Buti M, Dore GJ, et al. Once-daily simeprevir (TMC435) with pegylated interferon and ribavirin in treatment-naïve genotype 1 hepatitis C: the randomized PILLAR study. Hepatology December 2013;58(6):1918–29.

[45] Jacobson IM, Dore GJ, Foster GR, et al. Simeprevir with pegylated interferon alfa 2a plus ribavirin in treatment-naive patients with chronic hepatitis C virus genotype 1 infection (QUEST-1): a phase 3, randomised, double-blind, placebo-controlled trial. Lancet 2014;384:403–13.

[46] Manns M, Marcellin P, Poordad F, et al. Simeprevir with pegylated interferon alfa 2a or 2b plus ribavirin in treatment-naive patients with chronic hepatitis C virus genotype 1 infection (QUEST-2): a randomised, double-blind, placebo-controlled phase 3 trial. Lancet 2014;384:414–26.

[47] Zeuzem S, Berg T, Gane E, et al. Simeprevir increases rate of sustained virologic response among treatment-experienced patients with HCV genotype-1 infection: a phase IIb Trial. Gastroenterology 2014;146:430–41.

[48] Forns X, Lawitz E, Zeuzem S, et al. Simeprevir with peginterferon and ribavirin leads to high rates of SVR in patients with HCV genotype 1 who relapsed after previous therapy: a phase 3 trial. Gastroenterology 2014;146:1669–79.

[49] Moreno C, Hezode C, Marcellin P, et al. Efficacy and safety of simeprevir with PegIFN/ribavirin in naïve or experienced patients infected with chronic HCV genotype 4. J Hepatol 2015;62:1047–55.

[50] Everson G, Cooper C, Hézode C, Shiffman ML, Yoshida E, Beltran-Jaramillo T, Andreone P, Bruno S, Ferenci P, Zeuzem S, Brunda M, Le Pogam S, Nájera I, Zhou J, Navarro MT, Voulgari A, Shulman NS, Yetzer ES. DAUPHINE: a randomized phase II study of dano-previr/ritonavir plus peginterferon alpha-2a/ribavirin in HCV genotypes 1 or 4. Liver Int January 2015;35(1):108–19.

[51] Zeuzem S, Dufour JF, Buti M, Soriano V, Buynak RJ, Mantry P, Taunk J, Stern JO, Vinisko R, Gallivan JP, Böcher W, Mensa FJ. SOUND-C3 study group. Interferon-free treatment of chronic hepatitis C with faldaprevir, deleobuvir and ribavirin: SOUND-C3, a Phase 2b study. Liver Int February 2015;35(2):417–21.

[52] Sperl J, Horvath G, Halota W, Ruiz-Tapiador JA, Streinu-Cercel A, Jancoriene L, Werling K, Kileng H, Koklu S, Gerstoft J, Urbanek P, Flisiak R, Leiva R, Kazenaite E, Prinzing R, Patel S, Qiu J, Asante-Appiah E, Wahl J, Nguyen BY, Barr E, Platt HL. Efficacy and safety of elbasvir/grazoprevir and sofosbuvir/pegylated interferon/ribavirin: a phase III randomized controlled trial. J Hepatol 2016. S0168-8278(16)30429-9.

[53] Zeuzem S, Ghalib R, Reddy KR, et al. Grazoprevir-elbasvir combination therapy for treat-ment-naive cirrhotic and noncirrhotic patients with chronic hepatitis C virus genotype 1, 4, or 6 infection: a randomized trial. Ann Intern Med 2015;163(1):1–13.

[54] Dore G, Altice F, Litwin AH, et al. C-EDGE CO-STAR: efficacy of grazoprevir and elbasvir in persons who inject drugs (PWID) receiving opioid agonist therapy. In: Presented at the 2015 Annual Meeting of the American Association for the Study of Liver Diseases, San Francisco; November 13–17. 2015.

[55] Lawitz E, Gane E, Pearlman B, et al. Efficacy and safety of 12 weeks versus 18 weeks of treatment with grazoprevir (MK-5172) and elbasvir (MK-8742) with or without ribavirin for hepatitis C virus genotype 1 infection in previously untreated patients with cirrhosis and patients with previous null response with or without cirrhosis (C-WORTHY): a randomised, open-label phase 2 trial. Lancet 2015;385:1075–86.

[56] Buti M, Gordon SC, Zuckerman E, Lawitz E, Calleja JL, Hofer H, Gilbert C, Palcza J, Howe AY, DiNubile MJ, Robertson MN, Wahl J, Barr E, Forns X. Grazoprevir, elbasvir, and ribavirin for chronic hepatitis C virus genotype 1 infection after failure of pegylated inter-feron and ribavirin with an earlier-generation protease inhibitor: final 24-week results from C-SALVAGE. Clin Infect Dis 2016;62(1):32–6.

[57] Rockstroh JK, Nelson M, Katlama C, et al. Efficacy and safety of grazoprevir (MK-5172) and elbasvir (MK-8742) in patients with hepatitis C virus and HIV co-infection (C-EDGE CO-INFECTION): a non-randomised, open-label trial. Lancet HIV 2015;2:e319–27.

[58] Soriano V, Vispo E, de Mendoza C, Labarga P, Fernandez-Montero JV, Poveda E, Treviño A, Barreiro P. Hepatitis C therapy with HCV NS5B polymerase inhibitors. Expert Opin Pharmacother 2013;14(9):1161–70.

[59] Marascio N, Torti C, Liberto M, Focà A. Update on different aspects of HCV variability: focus on NS5B polymerase. BMC Infect Dis 2014;14(Suppl. 5):S1.

[60] Bhatia HK, Singh H, Grewal N, Natt NK. Sofosbuvir: a novel treatment option for chronic hepatitis C infection. J Pharmacol Pharmacother 2014;5(4):278–84.

[61] Kowdley KV, Lawitz E, Crespo I, et al. Sofosbuvir with pegylated interferon alfa-2a and ribavirin for treatment-naive patients with hepatitis C genotype-1 infection (ATOMIC): an open-label, randomised, multicentre phase 2 trial. Lancet June 15, 2013;381(9883):2100–7.

[62] Gane EJ, Stedman CA, Hyland RH, Ding X, Svarovskaia E, Symonds WT, Hindes RG, Berrey MM. Nucleotide polymerase inhibitor sofosbuvir plus ribavirin for hepatitis C. N Engl J Med January 3, 2013;368(1):34–44.

[63] Foster GR, Pianko S, Brown A, et al. Efficacy of sofosbuvir plus ribavirin with or without peginterferon-alfa in patients with hepatitis C virus genotype 3 infection and treatment-experienced patients with cirrhosis and hepatitis C virus genotype 2 infection. Gastroenterology 2015;149:1462–70.

[64] Kwo P, Gitlin N, Nahass R, Bernstein D, Etzkorn K, Rojter S, Schiff E, Davis M, Ruane P, Younes Z, Kalmeijer R, Sinha R, Peeters M, Lenz O, Fevery B, De La Rosa G, Scott J, Witek J. Simeprevir plus sofosbuvir (12 and 8 weeks) in hepatitis C virus genotype 1-infected patients without cirrhosis: OPTIMIST-1, a phase 3, randomized study. Hepatology August 2016;64(2):370–80.

[65] Lawitz E, Matusow G, DeJesus E, Yoshida EM, Felizarta F, Ghalib R, Godofsky E, Herring RW, Poleynard G, Sheikh A, Tobias H, Kugelmas M, Kalmeijer R, Peeters M, Lenz O, Fevery B, De La Rosa G, Scott J, Sinha R, Witek J. Simeprevir plus sofosbuvir in patients with chronic hepatitis C virus genotype 1 infection and cirrhosis: a phase 3 study (OPTIMIST-2). Hepatology August 2016;64(2):360–9.

[66] Sulkowski MS, Vargas HE, Di Bisceglie AM, Kuo A, Reddy KR, Lim JK, Morelli G, Darling JM, Feld JJ, Brown RS, Frazier LM, Stewart TG, Fried MW, Nelson DR, Jacobson IM, HCV-Target Study Group. Effectiveness of simeprevir plus sofosbuvir, with or without ribavirin, in real-world patients with HCV genotype 1 infection. Gastroenterology February 2016;150(2):419–29.

[67] Gane EJ, Stedman CA, Hyland RH, et al. Efficacy of nucleotide polymerase inhibitor sofosbuvir plus the NS5A inhibitor ledipasvir or the NS5B non-nucleoside inhibitor GS-9669 against HCV genotype 1 infection. Gastroenterology 2013. S0016-5085(13)01653-3.

[68] Kohli A, Osinusi A, Sims Z, et al. Virological response after 6 week triple-drug regimens for hepatitis C: a proof-of-concept phase 2A cohort study. Lancet 2015;385:1107.

[69] Lawitz E, Poordad FF, Pang PS, Hyland RH, Ding X, Mo H, Symonds WT, McHutchison JG, Membreno FE. Sofosbuvir and ledipasvir fixed-dose combination with and without ribavirin in treatment-naive and previously treated patients with genotype 1 hepatitis C virus infection (LONESTAR): an open-label, randomised, phase 2 trial. Lancet February 8, 2014;383(9916):515–23.

[70] Backus LI, Belperio PS, Shahoumian TA, Loomis TP, Mole LA. Real-world effectiveness of ledipasvir/sofosbuvir in 4,365 treatment-naive, genotype 1 hepatitis C-infected patients. Hepatology 2016;64(2):405–14.

[71] Gane EJ, Hyland RH, An D, et al. Efficacy of ledipasvir and sofosbuvir, with or without ribavirin, for 12 weeks in patients with HCV genotype 3 or 6 infection. Gastroenterology 2015;149:1454–61.

[72] Kohli A, Kapoor R, Sims Z, et al. Ledipasvir and sofosbuvir for hepatitis C genotype 4: a proof-of-concept, single-centre, open-label phase 2a cohort study. Lancet Infect Dis 2015;15:1049–54.

[73] Manns M, Samuel D, Gane EJ, Mutimer D, McCaughan G, Buti M, Prieto M, Calleja JL, Peck-Radosavljevic M, Müllhaupt B, Agarwal K, Angus P, Yoshida EM, Colombo M, Rizzetto M, Dvory-Sobol H, Denning J, Arterburn S, Pang PS, Brainard D, McHutchison JG, Dufour JF, Van Vlierberghe H, van Hoek B, Forns X. SOLAR-2 investigators. Ledipasvir and sofosbuvir plus ribavirin in patients with genotype 1 or 4 hepatitis C virus infection and advanced liver disease: a multicentre, open-label, randomised, phase 2 trial. Lancet Infect Dis 2016;16(6):685–97.

[74] Abergel A, Asselah T, Metivier S, et al. Ledipasvir-sofosbuvir in patients with hepatitis C virus genotype 5 infection: an open-label, multicentre, single-arm, phase 2 study. Lancet Infect Dis 2016;16:459–64.

[75] Osinusi A, Townsend K, Kohli A, et al. Virologic response following combined ledipasvir and sofosbuvir administration in patients with HCV genotype 1 and HIV co-infection. JAMA 2015;313:1232–9.

[76] Ingiliz P, Christensen S, Kimhofer T, Hueppe D, Lutz T, Schewe K, Busch H, Schmutz G, Wehmeyer MH, Boesecke C, Simon KG, Berger F, Rockstroh JK, Schulze Zur Wiesch J, Baumgarten A, Mauss S. Sofosbuvir and Ledipasvir for 8 weeks for the treatment of chronic hepatitis C virus infection in HCV-mono-infected and HIV-HCV co-infected individuals - results from the German hepatitis C cohort (GECCO-01). Clin Infect Dis August 17, 2016.

[77] Rosenthal ES, Kottilil S, Polis MA. Sofosbuvir and ledipasvir for HIV/HCV co-infected patients. Expert Opin Pharmacother 2016;17(5):743–9.

[78] Osinusi A, Kohli A, Marti MM, et al. Re-treatment of chronic hepatitis C virus genotype 1 infection after relapse: an open-label pilot study. Ann Intern Med 2014;161:634–8.

[79] Chahine EB, Sucher AJ, Hemstreet BA. Sofosbuvir/velpatasvir: the first pangenotypic direct-acting antiviral combination for hepatitis C. Ann Pharmacother September 8, 2016.

[80] Lawitz E, Reau N, Hinestrosa F, Rabinovitz M, Schiff E, Sheikh A, Younes Z, Herring Jr R, Reddy KR, Tran T, Bennett M, Nahass R, Yang JC, Lu S, Dvory-Sobol H, Stamm LM, Brainard DM, McHutchison JG, Pearlman B, Shiffman M, Hawkins T, Curry M, Jacobson I. Efficacy of sofosbuvir, velpatasvir, and GS-9857 in patients with genotype 1 hepatitis C virus infection in an open-label, phase 2 trial. Gastroenterology July 30, 2016.

[81] Feld JJ, Jacobson IM, Hézode C, et al. Sofosbuvir and velpatasvir for hcv genotype 1, 2, 4, 5, and 6 infection. N Engl J Med 2015;373:2599–607.

[82] Foster GR, Afdhal N, Roberts SK, et al. Sofosbuvir and velpatasvir for HCV genotype 2 and 3 infection. N Engl J Med 2015;373:2608–17.

[83] Gane E, Kowdley KV, Pound D, Stedman CA, Davis M, Etzkorn K, Gordon SC, Bernstein D, Everson G, Rodriguez-Torres M, Tsai N, Khalid O, Yang JC, Lu S, Dvory-Sobol H, Stamm LM, Brainard DM, McHutchison JG, Tong M, Chung RT, Beavers K, Poulos JE, Kwo PY, Nguyen MH. Efficacy of sofosbuvir, velpatasvir, and GS-9857 in patients with HCV genotype 2, 3, 4, or 6 infections in an open-label, phase 2 trial. Gastroenterology July 30, 2016.

[84] Curry MP, O'Leary JG, Bzowej N, et al. Sofosbuvir and velpatasvir for hcv in patients with decompensated cirrhosis. N Engl J Med 2015;373:2618–28.

[85] Wyles D, Brau N, Kottilil S, et al. Sofosbuvir/velpatasvir fixed dose combination for 12 weeks in patients co-infected with HCV and HIV-1: the phase 3 ASTRAL-5 study. In: Presented at the 51st annual meeting of the European association for the study of the liver, Barcelona, April 13–17. 2016.

[86] Gao M, Nettles RE, Belema M, Snyder LB, Nguyen VN, Fridell RA, et al. Chemical genetics strategy identifies an HCV NS5A inhibitor with a potent clinical effect. Nature 2010;465:96–100.

[87] Schinazi R, Halfon P, Marcellin P, Asselah T. HCV direct-acting antiviral agents: the best interferon-free combinations. Liver Int 2014;34:69–78.

[88] Gao M, O'Boyle 2nd DR, Roberts S. HCV NS5A replication complex inhibitors. Curr Opin Pharmacol September 16, 2016;30:151–7.

[89] Sulkowski MS, Gardiner DF, Rodriguez-Torres M, et al. Daclatasvir plus sofosbuvir for previously treated or untreated chronic HCV infection. N Engl J Med 2014;370:211–21.

[90] Nelson DR, Cooper JN, Lalezari JP, et al. All-oral 12-week treatment with daclatasvir plus sofosbuvir in patients with hepatitis C virus genotype 3 infection: ALLY-3 phase III study. Hepatology 2015;61:1127–35.

[91] Poordad F, Schiff ER, Vierling JM, et al. Daclatasvir with sofosbuvir and ribavirin for hcv infection with advanced cirrhosis or post-liver transplant recurrence. Hepatology January 11, 2016.

[92] Wyles DL, Ruane PJ, Sulkowski MS, et al. Daclatasvir plus sofosbuvir for HCV in patients coinfected with HIV-1. N Engl J Med 2015;373:714–25.

[93] DAKLINZA ™ [package insert]. Bristol-Myers Squibb Corp; 2016.

[94] Lontok E, Harrington P, Howe A, et al. Hepatitis C virus drug resistance-associated substitutions: state of the art summary. Hepatology 2015;62:1623–32.

[95] American Association for the Study of Liver Diseases (AASLD) and Infectious Diseases Society of America (IDSA). Recommendations for testing, managing, and treating hepatitis C. http://www.hcvguidelines.org/full-report-view.

[96] Harvoni® [package insert]. Foster City (CA): Gilead Sciences, Inc; 2015.

[97] Deeks ED. Ombitasvir/paritaprevir/ritonavir plus dasabuvir: a review in chronic hcv genotype 1 infection. Drugs June 2015;75(9):1027–38.

[98] EASL recommendations on treatment of hepatitis C 2016. [Internet]. Available: http://www.easl.eu/medias/cpg/HCV2016/English-report.pdf.

[99] Wyles DL, Sulkowski MS, Dieterich D. Management of hepatitis C/HIV coinfection in the era of highly effective hepatitis C virus direct-acting antiviral therapy.. Clin Infect Dis July 15, 2016;63(Suppl 1):S3–11.

[100] Foster GR, Irving WL, Cheung MC, Walker AJ, Hudson BE, Verma S, McLauchlan J, Mutimer DJ, Brown A, Gelson WT, MacDonald DC, Agarwal K, Research HCV. Impact of direct acting antiviral therapy in patients with chronic hepatitis C and decompensated cirrhosis. J Hepatol June 2016;64(6):1224–31.

[101] Cotte L, Pugliese P, Valantin MA, Cuzin L, Billaud E, Duvivier C, Naqvi A, Cheret A, Rey D, Pradat P, Poizot-Martin I. Dat'AIDS study Group. Hepatitis C treatment initiation in HIV-HCV coinfected patients. BMC Infect Dis July 22, 2016;16:345.

[102] Gane EG, Agarwal K. Directly acting antivirals (DAAs) for the treatment of chronic hepatitis C virus infection in liver transplant patients: "A Flood of opportunity". Am J Transplant 2014;14:994–1002.

[103] Khatri A, Dutta S, Marbury TC, Preston RA, Rodrigues Jr L, Wang H, Awni WM, Menon RM. Pharmacokinetics and tolerability of anti-hepatitis C virus treatment with ombitasvir, paritaprevir, ritonavir, with or without dasabuvir, in subjects with renal impairment. Clin Pharmacokinet July 7, 2016.

[104] Polepally AR, Badri PS, Eckert D, Mensing S, Menon RM. Effects of mild and moderate renal impairment on ombitasvir, paritaprevir, ritonavir, dasabuvir, and ribavirin pharmacokinetics in patients with chronic hcv infection. Eur J Drug Metab Pharmacokinet 2016 May 10.

[105] Smolders EJ, de Kanter CT, van Hoek B, Arends JE, Drenth JP, Burger DM. Pharmacokinetics, efficacy, and safety of hepatitis C virus drugs in patients with liver and/or renal impairment. Drug Saf July 2016;39(7):589–611.

[106] Sorbera MA, Friedman ML, Cope R. New and emerging evidence on the use of second-generation direct acting antivirals for the treatment of hepatitis C virus in renal impairment. J Pharm Pract February 22, 2016.

[107] Wilson EM, Kattakuzhy S, Sidharthan S, Sims Z, Tang L, McLaughlin M, Price A, Nelson A, Silk R, Gross C, Akoth E, Mo H, Subramanian GM, Pang PS, McHutchison JG, Osinusi A, Masur H, Kohli A, Kottilil S. Successful retreatment of chronic hcv Genotype-1 infection with ledipasvir and sofosbuvir after initial short course therapy with direct-acting antiviral regimens. Clin Infect Dis February 1, 2016;62(3):280–8.

[108] Ray K. Therapy: retreatment of HCV infection in DAA nonresponders. Nat Rev Gastroenterol Hepatol May 2015;12(5):252.

[109] European Association for the Study of the Liver. EASL clinical practice guidelines: management of chronic hepatitis B virus infection. J Hepatol 2012;57:167–85.

[110] Liaw YF, Kao JH, Piratvisuth T, et al. Asian-Pacific consensus statement on the management of chronic hepatitis B: a 2012 update. Hepatol Int 2012;6:531–61.

[111] Potthoff A, Berg T, Wedemeyer H. Late hepatitis B virus relapse in patients coinfected with hepatitis B virus and hepatitis C virus after antiviral treatment with pegylated interferon-a2b and ribavirin. Scand J Gastroenterol 2009;44:1487–90.

[112] Pawlotsky JM. Hepatitis C virus resistance to direct-acting antiviral drugs in interferon-free regimens. Gastroenterology 2016;151(1):70–86.

[113] Thomson E, Ip CL, Badhan A, Christiansen MT, Adamson W, Ansari MA, Bibby D, Breuer J, Brown A, Bowden R, Bryant J, Bonsall D, Da Silva Filipe A, Hinds C, Hudson E, Klenerman P, Lythgow K, Mbisa JL, McLauchlan J, Myers R, Piazza P, Roy S, Trebes A, Sreenu VB, Witteveldt J, STOP-HCV Consortium, Barnes E, Simmonds P. Comparison of next-generation sequencing technologies for comprehensive assessment of full-length hepatitis C viral genomes. J Clin Microbiol October 2016;54(10):2470–84.

[114] Forns X, Bukh J, Purcell RH. The challenge of developing a vaccine against hepatitis C virus. J Hepatol 2002;37:684–95.

[115] Reyes-Sandoval A, Ertl HC. DNA vaccines. J Curr Mol Med 2001;1:217–43.

[116] Houghton M, Abrignani S. Prospects for a vaccine against the hepatitis C virus. Nature 2005;436:961–6.

[117] Chattergoon M, Boyer J, Weiner DB. Genetic immunization: a new era in vaccines and immune therapeutics. FASEB J 1997;11:753–63.

[118] Siler CA, McGettigan JP, Dietzschold B, et al. Live and killed rhabdovirus-based vectors as potential hepatitis C vaccines. Virology 2002;292:24–34.

[119] Chi-Tan H. Vaccine development for hepatitis C: lessons from the past turn into promise for the future. Tzu Chi Med J 2005;17:61–74.

[120] Choo QL, Kuo G, Ralston R, et al. Vaccination of chimpanzees against infection by the hepatitis C virus. Proc Natl Acad Sci USA 1994;91:1294–8.

[121] Tang LL, Liu KZ. Recent advances in DNA vaccine of hepatitis virus. Hepatobiliary Pancreat Dis Int 2002;1:228–31.

[122] Alekseeva E, Sominskaya I, Skrastina D, et al. Enhancement of the expression of HCV core gene does not enhance core-specific immune response in DNA immunization: advantages of the heterologous DNA prime, protein boost immunization regimen. Genet Vaccines Ther 2009;7:7.

[123] Sarobe P, Lasarte JJ, Zabaleta A, et al. Hepatitis C virus structural proteins impair dendritic cell maturation and inhibit in vivo induction of cellular immune responses. J Virol 2003;77:10862–71.

[124] Large MK, Kittlesen DJ, Hahn YS. Suppression of host immune response by the core protein of hepatitis C virus: possible implications for hepatitis C virus persistence. J Immunol 1999;162:931–8.

FURTHER READING

[1] Keating GM. Elbasvir/grazoprevir: first global approval. Drugs April 2016;76(5):617–24.

Appendix

WORLD BANK LIST OF COUNTRIES ACCORDING TO ECONOMY

For the current 2017 fiscal year, low-income economies are defined as those with a gross national income (GNI) per capita, calculated using the *World Bank Atlas method*, of $1025 or less in 2015; lower middle-income economies are those with a GNI per capita between $1026 and $4035; upper middle-income economies are those with a GNI per capita between $4036 and $12,475; and high-income economies are those with a GNI per capita of $12,476 or more.

By Region
East Asia and Pacific
Europe and Central Asia
Latin America and the Caribbean
Middle East and North Africa
North America
South Asia
Sub-Saharan Africa

By Income
Low-income economies
Lower middle-income economies
Upper middle-income economies
High-income economies

East Asia and Pacific (38)

American Samoa	Korea, Republic of	Philippines
Australia	Lao PDR	Samoa
Brunei Darussalam	Macao SAR, China	Singapore
Cambodia	Malaysia	Solomon Islands
China	Marshall Islands	Taiwan, China
Fiji	Micronesia, Federal States of	Thailand
French Polynesia	Mongolia	Timor-Leste
Guam	Myanmar	Papua New Guinea
Hong Kong SAR, China	**Nauru**	Tonga
Indonesia	New Caledonia	Tuvalu
Japan	New Zealand	Vanuatu
Kiribati	Northern Mariana Islands	Vietnam
Korea, Democratic People's Republic of	Palau	

Europe and Central Asia (58)

Albania	**Gibraltar**	Norway
Andorra	Greece	Poland
Armenia	Greenland	Portugal
Austria	Hungary	Romania
Azerbaijan	Iceland	Russian Federation
Belarus	Ireland	San Marino
Belgium	Isle of Man	Serbia
Bosnia and Herzegovina	Italy	Slovak Republic

Bulgaria
Channel Islands
Croatia
Cyprus
Czech Republic
Denmark
Estonia
Faroe Islands
Finland
France
Georgia
Germany

Kazakhstan
Kosovo
Kyrgyz Republic
Latvia
Liechtenstein
Lithuania
Luxembourg
Macedonia, FYR
Moldova
Monaco
Montenegro
Netherlands

Slovenia
Spain
Sweden
Switzerland
Tajikistan
Turkey
Turkmenistan
Ukraine
United Kingdom
Uzbekistan

Latin America and the Caribbean (42)

Antigua and Barbuda
Argentina
Aruba
Bahamas, The
Barbados
Belize
Bolivia
Brazil

British Virgin Islands
Cayman Islands
Chile
Colombia
Costa Rica
Cuba

Curacao
Dominica
Dominican Republic
Ecuador
El Salvador
Grenada
Guatemala
Guyana

Haiti
Honduras
Jamaica
Mexico
Nicaragua
Panama

Paraguay
Peru
Puerto Rico
Sint Maarten (Dutch part)
St. Kitts and Nevis
St. Lucia
St. Martin (French part)
St. Vincent and the
Grenadines
Suriname
Trinidad and Tobago
Turks and Caicos Islands
Uruguay
Venezuela, RB
Virgin Islands (USA)

Middle East and North Africa (21)

Algeria
Bahrain
Djibouti
Egypt, Arab Republic of
Iran, Islamic Republic of
Iraq
Israel

Jordan
Kuwait
Lebanon
Libya
Malta
Morocco
Oman

Qatar
Saudi Arabia
Syrian Arab Republic
Tunisia
United Arab Emirates
West Bank and Gaza
Yemen, Republic of

North America (3)

Bermuda

Canada

United States

South Asia (8)

Afghanistan
Bangladesh
Bhutan

India
Maldives
Nepal

Pakistan
Sri Lanka

Sub-Saharan Africa (48)

Angola
Benin
Botswana

Gabon
Gambia, The
Ghana

Nigeria
Rwanda
São Tomé and Principe

Burkina Faso
Burundi
Cabo Verde
Cameroon
Central African Republic
Chad
Comoros
Congo, Democratic
Republic of
Congo, Republic of
Côte d'Ivoire
Equatorial Guinea
Eritrea
Ethiopia

Guinea
Guinea-Bissau
Kenya
Lesotho
Liberia
Madagascar
Malawi
Mali

Mauritania
Mauritius
Mozambique
Namibia
Niger

Senegal
Seychelles
Sierra Leone
Somalia
South Africa
South Sudan
Sudan
Swaziland

Tanzania
Togo
Uganda
Zambia
Zimbabwe

Low-Income Economies ($1025 or Less) (31)

Afghanistan
Benin
Burkina Faso
Burundi

Central African Republic
Chad
Comoros
Congo, Democratic
Republic of
Eritrea
Ethiopia
Gambia, The

Guinea
Guinea-Bissau
Haiti
Korea, Democratic People's
Republic of
Liberia
Madagascar
Malawi
Mali

Mozambique
Nepal
Niger

Rwanda
Senegal
Sierra Leone
Somalia

South Sudan
Tanzania
Togo
Uganda

Zimbabwe

Lower Middle-Income Economies ($1026–$4035) (52)

Armenia
Bangladesh
Bhutan
Bolivia
Cabo Verde
Cambodia
Cameroon

Congo, Republic of
Côte d'Ivoire
Djibouti
Egypt, Arab Republic of
El Salvador
Ghana
Guatemala
Honduras
India
Indonesia
Kenya

Kiribati
Kosovo
Kyrgyz Republic
Lao PDR
Lesotho
Mauritania
Micronesia, Federal
States of
Moldova
Mongolia
Morocco
Myanmar
Nicaragua
Nigeria
Pakistan
Papua New Guinea
Philippines
Samoa
São Tomé and Principe

Solomon Islands
Sri Lanka
Sudan
Swaziland
Syrian Arab Republic
Tajikistan
Timor-Leste

Tonga
Tunisia
Ukraine
Uzbekistan
Vanuatu
Vietnam
West Bank and Gaza
Yemen, Republic of
Zambia

Upper Middle-Income Economies ($4036–$12,475) (56)

Albania	Ecuador	Montenegro
Algeria	Fiji	Namibia
American Samoa	Gabon	Palau
Angola	**Georgia**	Panama
Argentina	Grenada	Paraguay
Azerbaijan	**Guyana**	Peru
Belarus	Iran, Islamic Republic of	Romania
Belize	Iraq	**Russian Federation**
Bosnia and Herzegovina	Jamaica	Serbia
Botswana	Jordan	South Africa
Brazil	Kazakhstan	St. Lucia
Bulgaria	Lebanon	St. Vincent and the Grenadines
China	Libya	Suriname
Colombia	Macedonia, FYR	Thailand
Costa Rica	Malaysia	Turkey
Cuba	Maldives	Turkmenistan
Dominica	Marshall Islands	Tuvalu
Dominican Republic	Mauritius	**Venezuela, RB**
Equatorial Guinea	Mexico	

High-Income Economies ($12,476 or More) (79)

Andorra	**Gibraltar**	Oman
Antigua and Barbuda	Greece	Poland
Aruba	Greenland	Portugal
Australia	Guam	Puerto Rico
Austria	Hong Kong SAR, China	Qatar
Bahamas, The	Hungary	San Marino
Bahrain	Iceland	Saudi Arabia
Barbados	Ireland	Seychelles
Belgium	Isle of Man	Singapore
Bermuda	Israel	Sint Maarten (Dutch part)
British Virgin Islands	Italy	Slovak Republic
Brunei Darussalam	Japan	Slovenia
Canada	Korea, Republic of	Spain
Cayman Islands	Kuwait	St. Kitts and Nevis
Channel Islands	Latvia	St. Martin (French part)
Chile	Liechtenstein	Sweden
Croatia	Lithuania	Switzerland
Curaçao	Luxembourg	Taiwan, China
Cyprus	Macao SAR, China	Trinidad and Tobago
Czech Republic	Malta	Turks and Caicos Islands
Denmark	Monaco	United Arab Emirates
Estonia	**Nauru**	United Kingdom
Faroe Islands	Netherlands	United States
Finland	New Caledonia	Uruguay
France	New Zealand	Virgin Islands (U.S.)
French Polynesia	Northern Mariana Islands	
Germany	Norway	

IDA (60)

Afghanistan	Haiti	Niger
Bangladesh	Honduras	Rwanda
Benin	Kenya	Samoa
Bhutan	Kiribati	São Tomé and Principe
Burkina Faso	Kosovo	Senegal
Burundi	Kyrgyz Republic	Sierra Leone
Cambodia	Lao PDR	Solomon Islands
Central African Republic	Lesotho	Somalia
Chad	Liberia	South Sudan
Comoros	Madagascar	Sudan
Congo, Democratic Republic of	Malawi	**Syrian Arab Republic**
Côte d'Ivoire	Maldives	Tajikistan
Djibouti	Mali	Tanzania
Eritrea	Marshall Islands	Togo
Ethiopia	Mauritania	Tonga
Gambia, The	Micronesia, Federal States of	Tuvalu
Ghana	Mozambique	Uganda
Guinea	Myanmar	Vanuatu
Guinea-Bissau	Nepal	Yemen, Republic of
Guyana	Nicaragua	Zambia

Blend (18)

Bolivia	Moldova	St. Lucia
Cabo Verde	Mongolia	St. Vincent and the Grenadines
Cameroon	Nigeria	Timor-Leste
Congo, Republic of	Pakistan	Uzbekistan
Dominica	Papua New Guinea	Vietnam
Grenada	Sri Lanka	Zimbabwe

IBRD (66)

Albania	Equatorial Guinea	Palau
Algeria	Fiji	Panama
Angola	Gabon	Paraguay
Antigua and Barbuda	Georgia	Peru
Argentina	Guatemala	Philippines
Armenia	India	Poland
Azerbaijan	Indonesia	Romania
Belarus	Iran, Islamic Republic of	Russian Federation
Belize	Iraq	Serbia
Bosnia and Herzegovina	Jamaica	Seychelles
Botswana	Jordan	South Africa
Brazil	Kazakhstan	St. Kitts and Nevis
Bulgaria	Lebanon	Suriname
Chile	Libya	Swaziland
China	Macedonia, FYR	Thailand

Colombia
Costa Rica
Croatia
Dominican Republic
Ecuador
Egypt, Arab Republic of
El Salvador

Malaysia
Mauritius
Mexico
Montenegro
Morocco
Namibia
Nauru

Trinidad and Tobago
Tunisia
Turkey
Turkmenistan
Ukraine
Uruguay
Venezuela, RB

Index

Printed in the United States
By Bookmasters

Physician's Guide